Life Finds a Way

Also by Andreas Wagner

Arrival of the Fittest

Paradoxical Life

LIFE FINDS A WAY

What Evolution Teaches Us
About Creativity

Andreas Wagner

BASIC BOOKS
New York

Basic Books
Hachette Book Group
1290 Avenue of the Americas, New York, NY 10104
www.basicbooks.com

Printed in the United States of America

First Edition: June 2019

Published by Basic Books, an imprint of Perseus Books, LLC, a subsidiary of Hachette Book Group, Inc. The Basic Books name and logo is a trademark of the Hachette Book Group.

The Hachette Speakers Bureau provides a wide range of authors for speaking events. To find out more, go to www.hachettespeakersbureau.com or call (866) 376-6591.

The publisher is not responsible for websites (or their content) that are not owned by the publisher.

The Library of Congress has cataloged the hardcover edition as follows:
Names: Wagner, Andreas, 1967 January 26– author.
Title: Life finds a way : what evolution teaches us about creativity /
    Andreas Wagner.
Description: First edition. | New York : Basic Books, 2019. | Includes
    bibliographical references and index.
Identifiers: LCCN 2018046693 (print) | LCCN 2018047908 (ebook) |
    ISBN 9781541645356 (ebook) | ISBN 9781541645332 (hardcover)
Subjects: LCSH: Evolution (Biology)—Philosophy. | Creative ability.
Classification: LCC QH360.5 (ebook) | LCC QH360.5 .W34 2019 (print) |
    DDC 576.801—dc23
LC record available at https://lccn.loc.gov/2018046693

ISBNs: 978-1-5416-4533-2 (hardcover); 978-1-5416-4535-6 (ebook)

LSC-C

10 9 8 7 6 5 4 3 2 1

Contents

Prologue

Long before life itself arose, nature created not just swirling galaxies and the thermonuclear engines of suns. It also created glittering crystals, like the diamonds that take millions of years to gestate in the womb of our planet. And it created the complex organic molecules found in interstellar gases, meteorites, and deep-sea vents that would become the building blocks of life. Once these building blocks had assembled into the earliest living cells, Darwinian evolution kicked in. It taught life to sate its boundless hunger by harvesting energy from sunlight and from energy-rich molecules. Equipped with molecular power plants, life could then conquer every habitat of our planet, from open equatorial oceans to frigid Arctic ice shelves, from hot subsurface rocks to endless arid plains and ice-sheathed mountains.

As time passed, life's single cells assembled into specialized teams with thousands, millions, and eventually billions of members. These multicellular organisms evolved sensors that helped them navigate the world by smell, sound, and light. They learned to escape enemies and attack prey by burrowing, swimming, walking, and flying. And, eventually, their nervous systems evolved complex brains that could create and comprehend abstract symbols, such as those on this page. To them we owe the cave paintings of Lascaux and the landscapes of Monet, simple abacuses and complex supercomputers, Sumerian accounting tablets and James Joyce's *Ulysses*, Pythagoras's theorem and Schrödinger's equation.

As different as all these may appear, they are all products of nature's creativity, a phrase that might bring to mind the finches that use tools to scare insects from hideouts, or the chimpanzees that fashion primitive spears to hunt bush babies. But I mean a more universal form of creativity manifest in chemistry, biology, and culture.

Much of human creativity fits a definition widely used by psychologists: a creative idea or product is an original and appropriate solution to a problem.[1] Some problems are simple, such as how to hold a stack of paper together, and these problems often have simple solutions—staples or paper clips. Other problems are mind-bogglingly complex, such as how to beat humans in strategy board games like Go, and so are their solutions—artificial intelligences like AlphaGo. These examples are technological, but defining creativity as problem solving is useful in many other domains, including the arts. Yale University's George Kubler— a towering art historian of the twentieth century—said: "Every important work of art can be regarded...as a hard-won solution to some problem."[2] And that's more than just one man's opinion. We will later see that artificial intelligences can use problem-solving strategies to create artistic products, like stirring melodies. To be sure, today's artificial intelligences are not on a par with the greatest human creators, and perhaps no psychological definition of creativity will ever be able to capture a Mozart symphony, a Picasso painting, or a Rodin sculpture. But the psychological definition of creativity is still immensely useful because it covers a broad spectrum of human creative expression.

Even more important, it is useful far beyond human affairs because it applies to problems that life solved before brains like ours—or any brains at all—arose. An enzyme that cracks the chemical bonds of an energy-rich molecule is one solution to the problem of how to harvest energy. The optical marvels of eyes are solutions to the problem of how to escape predators or hunt prey. And the antifreeze proteins of cold-blooded animals are a solution to the problem of how to survive in subzero temperatures. Viewing creativity as problem solving is even relevant for problems that the universe solved long before life itself arose. A crystal, for example, is a solution to the problem of how to find a stable arrangement of atoms or molecules.

I am an evolutionary biologist, and my life's work is to understand the creative powers of biological evolution that are embodied in microscopic algae and giant redwood trees, in gut bacteria and African elephants. Every one of the millions of species alive today is the most recent link in a nearly endless chain of creative achievement that goes back all the way to life's origins. Every organism is the product of countless innovations, from the molecular machines inside its cells to the physical architecture of its body. They account for life-forms that move at lightning speed, are perfectly camouflaged, or are covered with solar panels. Life's overflowing creativity fascinates me to no end.

At my Zurich laboratory, a team of some twenty researchers and I study the DNA of diverse organisms to investigate how nature creates new forms of life and new kinds of molecules.

We also observe microbes in the laboratory over thousands of generations to study how they evolve to surmount seemingly insurmountable challenges. And we compare how their exploits resemble creative processes in other fields, including how crystals take shape, how molecules self-assemble, and how algorithms solve problems.

But I am not just a scientist. I am also a father and an educator, and I am looking for better ways to raise children, to educate the next generation of scientists, to hire the most creative researchers, and to build and sustain a team of them. These very practical problems have led me to explore a vast literature on human psychology, education research, organizational management, and the economics of innovation. In these explorations I have discovered astonishing similarities between natural and human creativity.

This book is about these similarities and much more. First, it is about things Charles Darwin did not know. His theory of evolution by natural selection was a monumental achievement, but it was only a beginning. One of the things Darwin did not—could not—know is that natural selection can face obstacles that it alone cannot overcome. This book explains what these obstacles are. And it explains the mechanisms of evolution that can overcome them.

Second, this book illustrates the similarities between human creativity and a modern, augmented view of Darwinian evolution. These similarities are not only numerous but also deep, as psychological, historical, and biological research will testify later in the book.

Third, and perhaps most important, this book explains how these similarities can help us solve many of the problems facing

humans today. They can help us raise children to live more fulfilling lives, enhance businesses innovation, and prepare entire
nations for a world where innovation drives global leadership.

One reason why creativity in nature and culture are similar is
that difficult problems—how to create a regular diamond lattice,
an energy-efficient predator, or a sensitive radio antenna—share
a fundamental property: they have myriad solutions—many of
them poor, some a bit better, fewer really good, and a rare few
that are superb—and we can think of these solutions as forming a
mountainous landscape, where poor solutions correspond to low
foothills and the best solutions form the highest peaks (Figure 1).

Such a landscape is called an adaptive landscape, a concept that originated with the Harvard-trained geneticist Sewall
Wright. In the early twentieth century, Wright performed breeding experiments at the US Department of Agriculture that aimed
to create superior cows, hogs, and sheep.[3] Through these experiments, Wright discovered something fundamental and odd: the
Darwinian recipe of selecting the best animals can fail to create
a superior breed. And he found out why. To explain his insights
to others, he created the concept of an adaptive landscape.

Figure 1.

Evolving populations of organisms—sharks that must conserve energy while hunting, bacteria that must disarm deadly antibiotics, or herbivores that need to survive on nutrient-poor leaves—explore solutions to the problems they face more or less blindly. Wright saw that such problem solving amounts to climbing a peak in an adaptive landscape. A Darwinian method to find a best solution would start with any solution, no matter how bad, and tinker with it, preserving only the steps that improved it. Preserving the good and discarding the bad is the essence of both natural selection and its closest human analog—competition among people and their organizations.[4] Natural selection is perfectly suited to conquering an adaptive landscape whenever that landscape has a single peak. Always driving uphill, selection will reliably find the highest peak of such a landscape. But in a more complex landscape with two peaks, a dozen peaks, hundreds of peaks, or too many peaks to count, natural selection is more than just imperfect. It can fail disastrously. Evolving organisms that need to get from one peak to the next-higher one—from one solution to a better one—need to traverse the valleys between them, and that's where natural selection comes up short. Because selection never accepts the worse for the better, it allows not a single downhill step and can therefore get stuck far below Mount Everest. The importance of this problem is hard to overstate. All of evolution's creative products—a million species and counting—are the end points of journeys through such landscapes. Natural selection is essential for this journey, but in a rugged, multipeaked landscape, it is not sufficient.

Sewall Wright discovered not only this problem with natural selection, but also one potential solution. It is a force of evolution called genetic drift. To understand what it achieves, consider a human analogy involving professional musicians, artists, or athletes whose performance has plateaued, and no matter how much they practice, they fail to rise above this plateau. Such professionals often need to deconstruct and relearn their most basic techniques. Champion golfer Tiger Woods did exactly that when he reconstructed his golf swing in 1997. He suffered a lackluster 1998 season, but broke new records in the years thereafter. Sometimes things need to get worse before they can get better.

Genetic drift allows life to do the same thing, and it is at least as important for evolution as natural selection is. An additional and separate mechanism, recombination, enables evolving organisms to make giant leaps through an adaptive landscape and helps them vault over obstacles on their way to the highest peaks. Recombination takes place wherever sex happens, be it the plain vanilla sex humans indulge in or the more bizarre and mysterious forms that enable even bacteria and plants to exchange genes with each other.

Landscapes have become a fundamental concept in modern science, one whose importance goes far beyond biology. Just as evolving organisms journey through adaptive landscapes, so too do bonding atoms and molecules journey through something called an energy landscape. Energy landscapes can be no less rugged than evolution's landscapes. Studying them reveals not just how nature creates sparkling diamonds and glittering snowflakes, but also how we can create better molecules to serve us.

The problems of computer science, whether routing air traffic at busy airports or playing the game Go, also have multiple solutions, which form what is known as a solution landscape. And computers can solve complex problems with the same methods that enabled life to evolve. Even more important, these methods can help computers produce creative works, including patentable electronic circuits and musical compositions that rival human compositions.

But even more intriguing than artificial intelligences are the problem solvers most familiar to us. These are our own minds, which navigate a mental landscape of possibilities with Darwinian processes similar to those that life, molecules, and algorithms use to explore their landscapes. Some of these creative journeys, like those of the painters Raphael and Paul Gauguin, take creators through different countries and continents, but many other journeys explore an inner realm. Among them is the journey described in 1891 by physicist and physician Hermann von Helmholtz about some problems he had solved in the theoretical physics of liquids:

> I had only succeeded in solving such problems after many devious ways . . . and by a series of fortunate guesses. I have to compare myself with an Alpine climber, who, not knowing the way, ascends slowly and with toil, and is often compelled to retrace his steps because his progress stopped; sometimes by reasoning, and sometimes by accident, he hits upon traces of a fresh path, which again leads him a little further; and finally, when he has reached the goal, he finds to his annoyance a royal road on which he might have ridden up if he had been clever enough to find the right starting-point at the outset.[5]

And so here we see, some thirty years before Wright published his work, Wright's own theory foreshadowed and applied to the human mind. Indeed, because our minds create using mechanisms not quite identical but similar to drift and

recombination, applying the lessons of biological evolution—
what I call landscape thinking—can help us make our individual
and collective minds work better. Landscape thinking can help
us improve how we think, how we raise our children, and how
we enhance innovation with the right schools and universities,
business policies, and governmental regulations. But landscape
thinking is also about more than maximizing innovation, productivity, or economic output. It shows us how creativity comes
from a single source: the ability to explore vast and complex
landscapes, a principle so fundamental that it applies wherever
the new, the useful, and the beautiful originate. Like all good
science, then, it shows us something profound about ourselves
and our world.

Chapter 1

The Cartography of Evolution

It was spring of 1915 when the German army first unleashed weaponized chlorine gas on Allied soldiers in World War I. That's when John Burdon Sanderson Haldane saved thousands of Allied lives by inhaling chlorine gas himself. J.B.S., as he was known, was a twenty-three-year-old officer who had trained in mathematics and the classics at Oxford University and was serving on the front lines in France when the gas attacks began. Unfortunately, the British army had issued ninety thousand useless gas masks to its troops. Together with his father, a physiologist at Oxford, J.B.S. was charged with developing more effective gas masks. They built themselves a small gas chamber, in which they breathed chlorine gas with and without gas masks until their lungs became "duly irritable."[1]

Self-experiments like this had a long tradition in the Haldane family. Haldane's father, who also inspected mines for the British government, had taught the young J.B.S. the effects of methane gas by letting him read Shakespeare aloud in a contaminated mine until he fainted. Later on, as a fellow at Oxford,

Haldane manipulated his blood's acidity by consuming hydro-chloric acid and other toxic chemicals in experiments that left him in serious pain, or with violent diarrhea, or panting for several days.[2]

But Haldane was much more than an oddball scientist with a penchant for self-experimentation. He was arguably the greatest polymath of his generation. A precocious child who learned to read before age three, Haldane was as well versed in the classics as he was in science and was described by a contemporary as possibly "the last man who might know all there was to be known."[3] In the sciences, he made discoveries in fields ranging from physiology and statistics to genetics, evolution, and biochemistry. Curiously, like other eminent creators we will encounter later, he was a bit myopic—not to say blind—when judging which of his breakthroughs was the most important. He thought it was a discovery about cytochrome oxidase—an enzyme important for respiration—but history issued a different verdict.[4]

Haldane is today best remembered for a body of mathematical work with singular importance for twentieth-century biology. Together with the English statistician Ronald Fisher and the American geneticist Sewall Wright, Haldane formed a triumvirate that turned evolutionary biology from a domain of naturalists like Darwin into an exact, mathematical science.

Darwin's key insight—that all life emerged from a common ancestor with a lot of help from natural selection—is well known.[5] Less well known is the sheer breadth of supportive evidence that his naturalist's mind amassed. This evidence includes the spectacular success of breeders, whose artificial selection brought forth attractive roses and productive wheat, as well as dogs as different as pugs and Rottweilers.[6] The evidence also includes an endless procession of ever-changing fossil forms, from traces of primitive worms in the most ancient of rocks to sophisticated invertebrates like ammonites and more recent life-forms like fishes, amphibians, reptiles, and eventually mammals. It includes the anatomy of animals as superficially different as rats and bats, whose skeletons are variants of the same blueprint and reveal their deep relatedness. It includes useless atavistic traits like the rudimentary eyes of fish whose ancestors took up residence in dark caves and the embryonic teeth of birds, which first grow and later melt away again—remnants of birds' toothy reptilian ancestors.

And Darwin's evidence includes the motley collections of species found on remote islands like Hawaii or the Galapagos—profuse in unusual birds, insects, and bats but impoverished in mammals and amphibians. This contrast is mystifying until one realizes that island faunas are not the febrile dream of a mad creator. Instead, they contain those continental species that can get there by wind or flight and are liberated from competition to radiate into a cornucopia of new forms.[7]

Darwin's theory emboldened naturalists to search for further evidence of evolution in action, and they did not have to search long before they found the intriguing case of the peppered moth *Biston betularia*. Like so many other organisms favored by biologists—for example, the tiny fruit fly *Drosophila* and the even tinier bacterium *Escherichia coli*—the peppered moth is anything but flashy. It is a perfectly inconspicuous inhabitant of our planet, and that's the point: it aims to fit in. The grainy salt-and-pepper speckles on its gray wings are perfect camouflage on the lichen-covered tree bark in its English habitat. The moth is possibly the most literal illustration of the term "survival of the fittest" that one could find.[8] Moths with speckled wings *fit* a tree's texture best—they are best adapted to the tree's surface—and thus stand a better chance of eluding the sharp eyes of predatory birds. Experiments that pin moths to a tree and monitor how often they get eaten by birds prove just that: in a forest of light-barked trees, darker moths get eaten more often. They are less fit, less well adapted.[9]

Dark moths arise from occasional DNA mutations that alter a gene affecting wing color. Such mutations create a new form—a new *allele*—of this gene, which helps turn the wing darker and exposes the moths to predators. The misfortune of dark moths changed after the start of the Industrial Revolution, when trees became increasingly covered in black soot that concealed dark moths and revealed light moths to predatory birds. Dark moths carrying this new allele were better adapted to polluted trees, and more of them survived bird attacks. As air pollution increased and covered more trees in soot, dark moths spread at the expense of their lighter brethren until they became the prevalent forms in polluted areas.

Together, short-lived moths, their large populations, and a quickly changing environment offered an opportunity for mathematically inclined scientists like Haldane. In the industrial town of Manchester, dark-winged moths had completely replaced light-winged moths within half a century. Knowing this, Haldane developed mathematical equations that allowed him to calculate how much more likely it was that a light moth would be eaten by a bird than a dark moth. The answer turned out to be about 30 percent.[10] This modest difference in fitness sufficed to transform an entire population's wing color within a human life span.[11]

The wing colors of the peppered moth are discrete variants, each caused by a different allele with a major effect on color. But most variation in nature is not like that. It is graded, continuous variation, like the many hues of green in the trees of a forest, the innumerable shades of brown in the coats of dogs, the wide-ranging sizes among different grains of wheat, and the extensive differences in human stature, from the famously short Pygmies to the famously tall Dutch people. This is *polygenic* variation, influenced by not just one but hundreds of genes, each with a tiny effect.

Here is where the second member of our triumvirate, Ronald Fisher, comes in. A Cambridge-trained mathematician, he helped father not only modern statistics, but also population genetics (and eight children, to boot). Fisher worked for ten years at Rothamstead, an agricultural research station. There, he analyzed data from plant breeders, which helped him extend Haldane's mathematical feat from discrete variation to such polygenic traits as height or yield. He demonstrated mathematically how strong selection must be—how many individuals must be culled from a herd of cows, what fraction of wheat plants should be allowed to survive—to predict how fast traits like milk yield and grain size could evolve in one generation. Not only was Fisher's work useful, its mathematical precision also made it a capstone to much of Darwin's work.

Sewall Wright, the third member of the triumvirate, worked in parallel to Fisher and Haldane. Like Fisher, Wright was tackling practical problems in agriculture, in his case about breeding the most productive cows, hogs, and sheep. But unlike the theoretician Fisher, Wright was not just mathematically adept, but also a dyed-in-the-wool experimentalist who performed breeding experiments on more than thirty thousand guinea pigs. (The milk yield of guinea pigs may interest no one, but they are vastly superior to cows for breeding experiments because they are smaller, reproduce faster, and can be kept in larger populations.) And during these experiments Wright noticed something odd: selecting the best animals for reproduction—Fisher's prescription for breeding success—when repeated over and over for multiple generations, did not always work well to create a superior breed. For example, during ongoing selection to improve one trait, like beef quality or milk yield, other traits often deteriorate, including two crucial ones: mortality and fertility. And when that happens, a breeder's greatest hope may have become just another evolutionary dead-end.

Wright also investigated more than a hundred years of pedigrees and records kept by animal breeders. All this data helped him see what the theoretician Fisher had missed: genes interact in mind-bogglingly complex ways. A gene that increases milk yield can reduce meat quality, another one that increases meat quality may reduce fertility, and a third one that increases fertility may also increase a cow's risk of dying from disease. And Wright's mathematical analysis taught him that these interactions are the reason why natural selection, although essential, need not be sufficient for evolution's progress.[12]

You might ask what guinea pigs and dairy cows could possibly teach us about how nature creates. The creative powers of animal breeding do indeed seem modest when we view breeds of cattle and varieties of corn against the millions of species in life's glorious diversity. But Darwin himself already reminded us in the *Origin of Species* of how much diversity human breeders have created in some species. A modern corn cob is barely recognizable as a descendant of its grass-like ancestor teosinte from Middle America, and Chihuahuas are so different from Great Danes that it stretches the imagination to call them members of the same species. The success of breeding is a microcosm of evolution's creative power, and it uses the same principles that evolution has employed for almost four billion years. This is why Wright's insights eventually helped us understand nature's creativity on a larger scale.

In 1932, Wright was invited to present his work at the Sixth International Congress of Genetics to a general audience of biologists. Unfortunately, the mathematics were beyond the average biologist's skill, and Wright needed to communicate his ideas in a more accessible way.[13]

This is how the fitness landscape was born.

A fitness landscape, also known as an adaptive landscape, is a visualization that allows us to picture evolution at work. It looks much like a topographic map of a mountain range, except that its axes—corresponding to the east–west and north–south dimensions of a map—describe different characteristics of an organism that can vary over a continuous range of values. Such characteristics might include a giraffe's height, a rose petal's color, or the wing coloration of the peppered moth, as shown on the horizontal axis in Figure 1.1. An organism at one location in the landscape has a specific trait value, such as a wing with a specific shade of gray. A DNA mutation that creates a different shade of gray moves the organism along one axis of the landscape. The vertical dimension in the landscape corresponds not to altitude but to the fitness that comes with the trait value. In the years before industrial soot soiled English forests, lighter moths fit the tree background better than darker moths, so they occupied higher elevations closer to the landscape's peak.

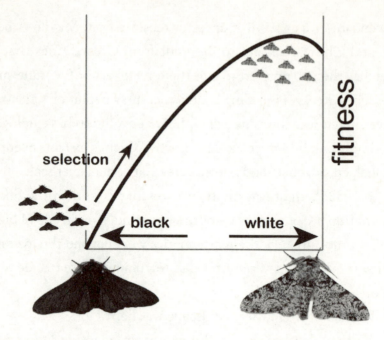

Figure 1.1.

Even a highly simplified two-dimensional landscape like that of Figure 1.1 already confers useful information. For example, the landscape has a single hump or "peak" close to the light end of the gray scale. All-black moths are easily picked off by birds, hence their fitness—on the far left—is far below peak. At the other extreme, snow-white moths also have some handicap because they are not perfectly matched to the mottled pattern of lichen-covered trees.

Over multiple generations, an evolving population of moths can be driven across this landscape by various evolutionary forces. One such driving force is DNA mutation, which creates new alleles. Mutation is blind, so a moth becomes either lighter or darker regardless of whether being lighter or darker would be the better move. A second force is natural selection—those moths who occupy the lowest parts of the slope farthest from the peak are most likely to get eaten by birds. Jointly, mutation and selection herd a population toward a peak, where most individuals are well adapted and resemble each other. Selection then keeps the population near the peak by culling those mutant outliers that are too far downslope.

As the environment changes, the locations of the peaks in a landscape can change, too. For example, the climate may become hostile to moths, or a new predator can appear, or pollution can transform lichen-covered trees into soot-covered trees. In this latter case, the landscape's peak shifts such that dark moths are preferred over light moths, as shown in Figure 1.2. The combined action of mutation and selection are still at work, but now they drive the population in the opposite direction, up the new peak.

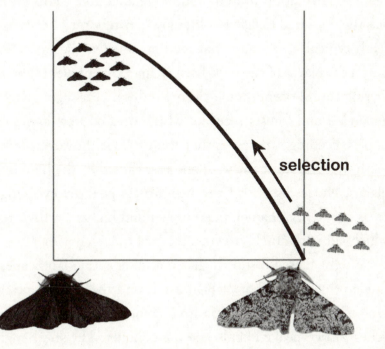

Figure 1.2.

This simple visualization of what natural selection does—
it drives a population up a landscape's peak—helped spread
Wright's ideas among biologists. Wright used the landscape as
a metaphor and was deliberately vague about the traits it could
represent.[14] That turned out to be fortuitous because it allowed
the landscape concept to become a veritable Rorschach inkblot
for evolutionary biologists, permitting ever-evolving interpreta-
tions of the basic idea. Among the first who realized its broad
explanatory power was the paleontologist George Gaylord Simpson, who used landscapes to describe evolutionary transforma-
tions that were more ancient and glacially slower than the recent
and swift evolution of the peppered moth. In his 1944 book,
Tempo and Mode in Evolution, Simpson illustrated the idea of a
fitness landscape with today's horses and their fifty-five-million-
year-long evolution from a diminutive ancestor.[15] This ancestor
was the dog-sized *Eohippus*—literally "dawn horse." *Eohippus* had
teeth typical of animals that feed on soft leaves, protected by
only a thin layer of the rock-hard enamel that prevents abrasion.
During the Miocene, some twenty million years ago, grasslands
expanded and forests receded, which created new habitats for
horses. Feeding on grass rather than foliage, however, requires
teeth that can resist the wear and tear caused by the harder grass
blades. Horses ascended the new fitness peak by evolving increasingly thick enamel, piled higher and higher on their teeth,
which led to the high-crowned teeth of today's horses.[16]

Wright had also shown that not all adaptive landscapes are
single-peaked like that of Figure 1.1, and that landscapes with
two or more peaks can arise from complex interactions among
genes. A two-peaked landscape was conquered by another group
of ancient organisms: the now extinct spiral-shaped mollusks

known as ammonites.[17] As an ammonite grew, it expanded its shell by adding material to the shell's growing rim, and it eventually secreted a wall—visible as a rib-like suture on the outer surface—to seal the shell's outermost part from its interior. Through multiple episodes of growth and wall building, the animal created a series of ever-larger sealed compartments that spiraled around a central axis (Figure 1.3). Unlike snail shells, ammonite shells were multi-chambered, but the animal inhabited only the outermost chamber. This chamber connected to the others through a siphuncle, a thin tube used to empty or fill these chambers, much like a submarine's ballast tanks, allowing the animal to rise toward the surface or descend into the depths of the ocean.

Although the soft parts of ammonites are rarely preserved, we can get an idea of how these animals propelled themselves through the water from a present-day relative, the nautilus. Ancestors of the nautilus discovered the principle of jet propulsion, which the nautilus still exploits, expelling water through a tube-like syphon near its mouth to push itself backward through the water.[18] Pushing your home through the water uses a lot of energy, and because energy is scarce out in the wild blue, it is

Figure 1.3.

important that a nautilus or an ammonite swims as efficiently as possible. A home with the right shape is crucial to achieving this efficiency.

Even though ammonites come in many sizes and shapes, the paleontologist David Raup realized in 1967 that these shapes could be categorized by two simple quantities. The first is the rate at which an ammonite increases its diameter while it grows and adds chambers, and the second is related to the diameter of the largest chamber opening, which is its gateway to the outside world.[19] A prototypical ammonite has the shape shown in the photograph on the left side of Figure 1.3, but other shapes also occur.[20] For example, ammonites that expand their diameter very slowly but have large chamber openings would resemble the one in the middle of Figure 1.3, whereas the opposite extreme—fast expansion and small opening—would correspond to the shape on the right side of the figure.

These two quantities are the two axes of a three-dimensional fitness landscape. The elevation reflects how easily an ammonite can propel itself through the ocean. John Chamberlain, a graduate student of Raup, was the first who measured this swimming efficiency.[21] He created dozens of Plexiglas models of various ammonite shapes and dragged them through a water tank to measure their drag coefficient, which is directly proportional to the amount of force needed to propel an animal through the water. The higher the drag coefficient, the more energy the animal needs in order to swim at a given speed.[22]

Chamberlain found that ammonites were ten times less efficient swimmers than those truly streamlined animals with an internal skeleton, like squid, fish, and dolphins.[23] That's the price they paid for being protected by a hard external skeleton. But swimming efficiency also varied among ammonites. This means that the three-dimensional fitness landscape of swimming efficiency is not flat. In fact, it turns out that the landscape has two peaks, a bit like the landscape shown in Figure 1.4.[24] That is, two distinct ammonite shapes are more efficient than all other shapes. The peaks corresponding to these shapes are separated by a valley of inferior shapes. If evolution has optimized ammonite shape for efficient swimming, then actual ammonite shapes should cluster near the peaks. Otherwise, they should be scattered haphazardly across the peaks and valleys.

To find out which was the case, Raup and others analyzed shape data from hundreds of ammonites, but they were in for a surprise. They found a third, unexpected possibility: the ammonites clustered around only one peak. The other one was mysteriously vacant. This could have happened if no mutations

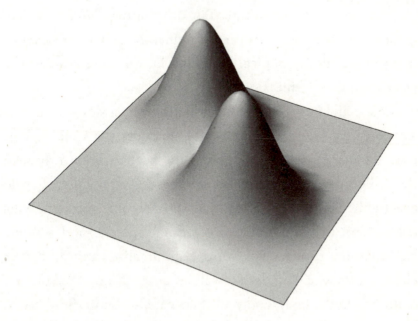

Figure 1.4.

had ever created ammonite shapes near the vacant peak. In that
case, natural selection would have had nothing to select, so the
peak would have remained unoccupied. But the actual solution
to this mystery was more mundane: a lack of data. By 2004, when
scientists had recorded the shapes of hundreds of additional ammonites, they found the second peak well occupied after all.[25]
Among all possible ammonite shapes, evolution favored the two
that swam most efficiently. In Wright's genetic language, the two
peaks would correspond to different combinations of genes that
helped create two different but equally optimal shapes for swimming. Sadly, we may never know which genes, nor how ammonites climbed those peaks, because they all died so many millions
of years ago.

The fitness landscapes of ammonites, horse teeth, and peppered moths are built on the hard foundation of physics—the
hydrodynamics of swimming, the mechanics of mastication,
and the optics of camouflage. But other fitness landscapes are
grounded in the softer realities of animal behavior, for example
in a genus of flashy tropical butterflies known as *Heliconius*, the
passion-vine butterflies.

One could be forgiven for wondering why a slow-flying, del-
icate creature like a butterfly would not adopt the same strategy
that has guaranteed the survival of the peppered moth through
the millennia: hide. That's because passion-vine butterflies do
exactly the opposite. Coming in a profusion of resplendent wing
colorations, they aim to show off. Some species sport a single
red stripe on a solid black wing, a minimalist pattern of sleek
elegance, some add a splash of yellow, others a fan of red rays ra-
diating from the body, and yet others parade a sunburst of bright
orange and yellow patches.

To see why an animal would advertise its presence with a flourish, it helps to know some other organisms that do the same. They include the gaudy but venomous coral snake and the dazzling but toxic poison dart frog. Their message could not be clearer: back off.

While passion-vine butterflies do not have dangerous fangs, they have a special trick to keep their enemies at bay: their larvae feed on passion vines, which produce dangerous self-defense chemicals, including cyanogenic glycosides. Butterfly larvae can tolerate these poisons, but once a butterfly larva has ingested them, the animal becomes toxic as well.[26]

Like billboards on a highway, which are most effective when you see them more than once, warning colors—the technical word is *aposematic colors*—are best remembered if many animals carry them. In other words, poisonous animals display strength in numbers. If many toxic butterflies in a patch of forest share the same color pattern, they reduce any one animal's risk of getting eaten. A naïve predator that survives biting a chunk out of one distasteful butterfly will remember that experience for life and avoid all others. But it will gladly take a bite out of a butterfly with a new color pattern, as experiments performed in 1972 by University of Washington zoologist Woodruff Benson prove. He painted over the red stripe on the wings of *Heliconius* butterflies with the wings' black background color. Sure enough, when he released the altered butterflies, he found that more of them than the originals were killed over time, and more of the survivors were mutilated, showing bite marks from predatory birds, reptiles, or mammals.[27]

With all of this in mind, imagine a fitness landscape whose two cardinal axes distinguish different color patterns on a butterfly wing. For example, one axis might quantify the amount of red, and the other the amount of yellow, relative to a black background. If many butterflies share a similar protective pattern, they create a peak in this landscape. Mutant butterflies whose colors lie off-peak are not protected and must run the gauntlet of hungry predators.

In the fitness landscape of warning coloration, a peak pulls evolving butterflies toward it because it guarantees safety in numbers. This pull turns out to be so great that even different species of passion-vine butterflies—distinguishable by their antennae, genitals, and other features—have evolved the same warning coloration.[28] They are all better off near the peak than anywhere else in the landscape. This is a remarkable example of convergent evolution, a process in which natural selection helps make different species more similar. It is also an example of Müllerian mimicry, the phenomenon where some toxic species mimic other toxic species, named after its discoverer, the nineteenth-century German naturalist Fritz Müller.

In contrast to a peppered moth, whose wing color needs to match that of a tree's bark, a butterfly's warning color is arbitrary, as long as many other butterflies share it and predators can recognize it. Nothing would prevent *Heliconius* butterflies in different geographical areas from showing different color patterns. In one population, all individuals might share that black wing with a single red stripe, whereas in another they might sport the sunburst of orange and yellow.

This is indeed the case, and in not just two but more than a dozen different areas, some covering thousands of square miles in the Amazon basin. What is more, different areas don't just harbor butterflies with different warning flags. Two species that mimic each other in one area often also mimic each other in another area. That would be less remarkable if the protective color patterns in the two areas were the same, because it could be explained if the species migrated between areas. But the color patterns in different areas can be completely different from one another. In other words, species in different areas have converged independently on their area's protective color pattern. Such multiple instances of convergent evolution underscore the power of the protection provided by a prevalent color pattern.[29]

We may never know with certainty how the geographic diversity of these warning color preferences originated, but a hint comes from the much cooler climate that existed on our planet in the Pleistocene starting some 2.5 million years ago. During this time, when large regions of the planet were sheathed in ice, the Amazonian forest habitats of *Heliconius* may have retreated to smaller forest islands separated by vast areas of open grasslands that could not be traversed by butterflies.[30] Imagine such isolated pockets of hospitable land as hothouses of evolution, where different butterfly populations could evolve different warning colorations. Once the globe warmed up again, these forest islands expanded into enormous and continuous swaths of rainforests. Butterfly populations expanded with them but were kept separate by natural barriers such as rivers and mountains.

Whatever the true origin of their different color patterns, the main take-home message is that the fitness landscape of passion vine butterfly coloration is not simple. It has multiple peaks, each corresponding to the warning color that dominates in a different region of the Amazon basin.[31]

When Sewall Wright conceived the fitness landscape, he did not have ammonites or butterflies in mind; he was thinking about his breeding experiments and the complex gene interactions they revealed. Wright's math showed that such interactions could bring forth fitness landscapes with many more than two or even a dozen peaks. More than that, he realized that their topography could be so complex that it defies imagination.

To see where Wright was coming from, let's revisit the peppered moth. While its wings can display many shades of gray, it turns out that a population of moths would mostly consist of two types: a light one referred to as *typica* and a dark one referred to as *carbonaria*.[32] In genetics jargon, these moths have two different *phenotypes*—a term that refers to any observable feature of an organism—and these phenotypes are encoded by two different *genotypes*, the DNA that is responsible for their appearance. The two genotypes are two different alleles of the same gene, which can be inherited in the indivisible, atom-like fashion that Gregor Mendel first discovered when he crossed pea plants in his monastery garden.[33] And because a moth's wing is basically either light or dark, one can replace the continuous light–dark axis of the one-dimensional landscape from Figure 1.2 with something simpler—the two points at the ends of the light–dark continuum, as shown in Figure 1.5a. Each of these points has a different value of fitness that describes how well a moth of that color can survive and reproduce. (The figure does not show that value.)

If it were only wing color that mattered to a moth's survival, the story would end here. But other traits also contribute, and that's where the complications begin. Wing size is one of these traits, and we know that mutations in some genes can alter it.

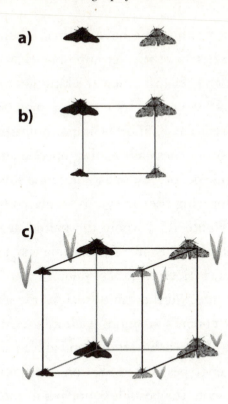

Figure 1.5.

Moths with one allele of such a gene would have normal large wings, whereas moths with the other, mutant allele would have smaller wings. Smaller wings reduce lift and impair flight and thus decrease fitness. Together with the two wing-color alleles, the two wing-size alleles can form four possible genotypes, which can be visualized as the four corners of the square shown in Figure 1.5b.

It gets still more complicated. Now consider a third gene, this one influencing the size of the moth's antennae. These marvelous sensory organs allow a male to home in on a female that is miles away and to follow a faint scent trail of just a handful of female pheromone molecules per cubic meter. Moths with one variant of this antenna gene have normal antennae, whereas moths with another have smaller antennae that are less sensitive, so they might get lost when tracking a female. Needless to say, getting lost while searching for your mate is not great for your ability to reproduce, another important aspect of fitness. With the addition of these alleles that encode antenna size we have eight possible genotypes: two for antenna size, two for wing color, and two for wing size. They can be placed on the corners of the cube in Figure 1.5c, where the paired, leaf-like objects stand for the antenna. (Just like Figures 1.5a and 1.5b, this figure does not show the genotypes' fitness values.)

Other genes affect further traits, such as the acuity of vision and the ability to endure starvation, evade attackers, or extract energy from nectar. With each new trait and allele pair that we add, the number of genotypes doubles. For one, two, and three traits, we were able to write the possible genotypes as the end points of a line, the corners of a square, and the vertices of a cube—objects in one, two, and three dimensions. But for four traits and their sixteen possible genotypes, we would need an object like a cube—but in four dimensions. Mathematicians call such high-dimensional cubes *hypercubes*. We can't visualize them well, but mathematics can describe them because their geometry follows straightforward laws. For example, the number of a hypercube's vertices doubles with every added dimension. A four-dimensional hypercube has sixteen vertices, a five-dimensional hypercube has thirty-two, a six-dimensional has sixty-four, and so on.

Despite the leading role that moths played in early evolutionary biology, they were soon surpassed by the tiny fruit fly *Drosophila melanogaster*. Geneticists cherish the fruit fly for several reasons: It is small, so one can easily keep thousands of flies in the lab. It's not a picky eater—a bit of yeast, cornmeal, or sugar, and happiness ensues. It reproduces very quickly. And despite its small size it has many traits that one can study with just a low-powered microscope, such as the shape of its wings, the color of its eyes, or the size of its antennae.

These advantages allowed geneticists like Thomas Hunt Morgan to scour thousands of fruit flies for mutant genes. Starting in 1908, Morgan toiled for two years before his first big break, when he discovered an allele he christened *white* because it turns the usually bright-red eyes of flies white.[34] After this first mutant allele was discovered, others quickly followed over the next years. They modified not only eye color, but all sorts of traits, including the size and shape of wings and body, the structure of important sensory organs like eyes, antennae, and bristles, as well as key traits like fertility and life expectancy.

By the time Wright proposed the fitness landscape concept in 1932, fruit fly experiments had already identified mutations in four hundred different fruit fly genes.[35] Even if each of these four hundred genes had only two alleles, there would be 2^{400}—or 10^{120}—possible genotypes, each with a fitness value potentially different from every other genotype. That's a very large number, much larger than the comparatively puny number of 10^{90} hydrogen atoms in the universe. As with the moths in Figure 1.5, each of these genotypes can still be placed on one vertex of a cube, albeit a four-hundred-dimensional one, with as many vertices as there are genotypes. The resulting landscape looks nothing like a familiar three-dimensional mountain range. Instead, each vertex of the cube corresponds to one "location" in the landscape—a fly with a specific allele combination—and its fitness is the "altitude" at this location.

This abstraction, far removed from our daily experience of a landscape, is what Wright had in mind when he introduced the landscape idea. But, like the rest of us, he could not visualize it with limited three-dimensional geometry. So he did what most of us do when faced with complexity far beyond our mind's grasp: he ignored it. He continued to talk about fitness landscapes as if they were three-dimensional and showed the kinds of peaks and valleys we are familiar with. And who can blame him? All our intuition about geometry comes from the three-dimensional world we live in. It may not apply to higher dimensions, but it is all we have.

Despite their limitations, even highly simplified landscapes and their peaks can be enormously valuable. Their topography can hold clues to how innovations emerge in biological evolution and how its creative process produced well-camouflaged moths, efficiently swimming ammonites, and gaudy poisonous butterflies. What's more, we shall see later that these landscapes are just as useful for understanding other forms of creativity. And even where the three-dimensional landscape metaphor fails—*especially* where it fails—it can teach us important lessons about creativity.

The complexity of evolution's landscapes also makes another point: when geneticists like Morgan and Wright first glimpsed life's genetic complexity, they got more than they had bargained for.

But, as it turns out, they had seen nothing yet.

Chapter 2

The Molecular Revolution

Morgan and his research associates—also known as the fly boys—discovered much more than the *white* gene. They also discovered that genes are located on chromosomes, a discovery that earned Morgan the 1933 Nobel Prize. And Morgan invented genetic mapping, which allowed scientists to locate genes like *white* on each of the five fruit fly chromosomes. Morgan's work still resonated half a century later, when his ideas helped locate in the human genome those genes involved in diseases like breast cancer. But an even bigger prize eluded him: to understand how different alleles of a gene cause different phenotypes to appear. That discovery had to wait for a molecular revolution in biology, which Morgan's work had prepared, but which would not get going until decades later.

It began when Oswald Avery showed in 1944 that DNA extracted from the corpse of a pneumonia bacterium can transform other, harmless bacteria into dangerous killers, as dangerous as live pneumonia bacteria. And it continued when, in 1953, James Watson and Francis Crick first elucidated the chemical structure of genotypes when they discovered DNA's double-stranded spiral staircase.[1] Each strand of this celebrated DNA double helix is built from four different nucleotide building blocks, distinguished by the four bases adenine, guanine, cytosine, and thymine and abbreviated by the letters A, C, G, and T that together form DNA's molecular alphabet. A molecule with this structure is an ideal information carrier because different sequences of the four letters, like different texts in the English language, can encode different information—the kind that parents pass on to their offspring.

When a cell decodes the information in a gene's DNA, it first transcribes the DNA's letter sequence into an RNA, or ribonucleic acid, copy. This RNA molecule is usually a mere intermediary. Its role is to be translated into a protein's letter sequence of amino acids. Once created, this amino acid string is incessantly jostled by the endless vibrations—also known as heat—of nearby molecules that collide with it. The energy in these collisions helps a protein fold into an intricate three-dimensional shape that biochemists call a conformation or a fold. The folded protein also vibrates with heat, and these vibrations allow proteins to perform myriad useful jobs. Protein enzymes catalyze thousands of different chemical reactions that take place in organisms on this planet, each enabled by an enzyme's unique three-dimensional shape. Proteins import hundreds of different nutrients into cells and help excrete just as many kinds of waste molecules. Proteins stiffen the molecular skeleton that prevents our cells from collapsing into amorphous blobs, and that makes brain cells visibly different from liver cells. Protein hormones keep our body working, like the insulin that controls our blood sugar, the prolactin that enables milk production, or various pain-reducing endorphins.[2] And proteins keep life on the move by spinning bacterial flagellae—themselves made of protein—and by contracting mammalian muscles. Without these work-horse molecules, life might never have crawled out of the primordial soup. All these proteins are encoded in the genes of the organisms that produce them. We humans have more than twenty thousand genes, an organism like a fruit fly has some fifteen thousand genes, and simpler organisms like the bacterium *Escherichia coli* still have a few thousand.[3]

Each of these genes can mutate anywhere along its DNA letter sequence. Such mutations occur when high-energy particles or atoms smash into DNA, when destructive by-products of metabolism react with DNA, or when DNA replication enzymes—another crucial kind of protein—commit errors while copying DNA. These processes create different kinds of mutations. One of them—also called a point mutation—is especially frequent and alters only a single letter in a gene. It is easy to calculate the total number of alleles that such molecular typos can create. For a gene with some one thousand nucleotide letters—not unusually long—the first letter must be one of the four possibilities: A, C, G, or T. Whichever it is, let's say a C, the letter can change into any of the three other letters (A, G, or T), so there are three mutant texts that can be created by altering the first letter. The same argument holds for the second of the thousand letters—it can change into three others—for the third letter, and so on, all the way through the one-thousandth letter. All of these possibilities add up to 3,000 new alleles that a single DNA typo can create. This number is even greater for longer genes, and for those mutations that alter more than one letter at a time.

All of this means that biologists today must reckon with landscapes of staggering complexity, much greater than Wright's landscapes with a few genes and a handful of alleles. If we just consider all the variants of a *Drosophila* genome that can be produced by a single typo in one of *Drosophila*'s fifteen thousand genes, the resulting landscape has a whopping 3000^{15000} possible genotypes.[4] That's a one with more than twelve thousand trailing zeroes, long enough to fill several pages of this book. All these genotypes can still be arranged on the corners of a high-dimensional cube, but the number of vertices on this cube is much greater than the number on the cube that Wright envisioned. Wright's hypercube already had more vertices than there are atoms in the universe. If each atom in our universe harbored another universe, and if each of these universes harbored as many atoms again as exist in our universe, the total number of atoms in all of these universes would still be dwarfed by the number of possible fruit fly genotypes.

The molecular revolution also clarified how exactly mutations change genotypes and phenotypes. A mutation that changes a single letter in a gene displaces a genotype on the vast hypercube of all DNA sequences, from one vertex to an adjacent vertex, and such changes often alter the encoded protein, which transforms the phenotype. The *white* gene, for example, affects the eye color of fruit flies not because it encodes an eye color pigment, but because it encodes a transporter protein that delivers the molecular building blocks of such eye pigments. Mutations in this gene cause white eyes because they cripple the transporter, such that building blocks never reach their destination in the eye.[5]

———

While the molecular revolution deepened our view of life immensely, it left the principle behind Wright's fitness landscapes untouched. Biologists still think of organisms as having some adaptive value or fitness. We still think of an organism's genotype as a location and its fitness as an elevation in a fitness landscape. And when populations of organisms find creative solutions to a problem they face, such as how to swim efficiently or how to escape predators, we still envision that they explore a fitness landscape and climb its peaks, and we still imagine this landscape as a mountain range in three dimensions because we cannot cram the vast hypercubes of high-dimensional genetic spaces into our limited minds.

Just like relentless competition enables some humans to win a race to the top, natural selection enables populations of organisms to climb any one peak in a fitness landscape. To do so, they need natural selection. But the complexity of the actual landscapes revealed by the molecular revolution lays bare the fact that natural selection is not enough. The landscape of ammonites had two peaks, and that of passion-vine butterflies had dozens of peaks, but truly complex landscapes can harbor many more. Not only can these peaks have different heights, but landscape topography can vary in many other ways as well. Some peaks might be gently sloped, whereas others might be jagged. The peaks may be isolated and scattered throughout the landscape, or they might form a continuous mountain range connected by ridges and saddles.

Landscapes like these can limit natural selection's powers because of selection's proverbial blindness. When selection works on a population scattered along the face of a mountain, it eliminates all downslope mutants and preserves only those upslope, blindly driving the population toward the nearest peak. A population that starts out at the foot of a low hill can get to the nearest peak—a *local* peak, in scientific jargon—but that's also where it will get stuck. In its relentless uphill drive, natural selection will not allow a population to cross the valley separating this local peak from the next higher peak. It will ruthlessly prevent inferior variants from surviving, just like those predators that kill butterflies with rare, off-peak warning colors. Even a population laboring to climb the slope of the very highest or global peak might reach some jagged outcrop on its way up. To rise further it would have to backtrack downhill at least a few steps before resuming its climb, but because of natural selection, it cannot do so. In rugged landscapes like that of Figure 1 in the prologue, the roof of the world might be near but forever out of reach because natural selection is like a powerful engine with a crucial flaw—it can only go uphill.

Wright was already worried that fitness landscapes have many peaks, but it was not until 1987 that it became clear how bad the problem can be.[6] That's when biologists Stuart Kauffman and Simon Levin estimated the number of peaks under the simplest possible theoretical assumption—that the fitness of different genotypes is drawn at random from some possible range of values. This assumption is as good a starting point as any other given how utterly impossible it would be to measure the fitness of all possible genotypes. Just consider the 10^{120} genotypes of Wright's original landscape. Even if every one of the seven billion humans alive today dropped everything they are doing and dedicated the next hundred years to the exceedingly important task of measuring the fitness of fruit flies, and did so at a speed of one fly per second, they would be able to process 10^{20} flies, a formidable number to be sure, but less than one in a 10^{100}th of all the flies in Wright's fruit fly landscape.[7]

The calculations of Kauffman and Levin show that even in their simplified, theoretical landscape, where every gene has only two possible alleles, about one in every fifteen thousand genotypes would be a peak. That number does not sound too bad, until you compute the total number of peaks. It turns out to be a one with more than four thousand trailing zeroes.[8] Thus, not only is the size of fitness landscapes beyond imagination, but the number of peaks can be no less mind-boggling. And to make matters worse, this number of peaks increases explosively with the number of possible genotypes.[9]

Only one among all these peaks is Mount Everest, the single global peak among myriad lower ones. Natural selection can fulfill its promise of finding the best-adapted organism only if Mount Everest can be reached through a constantly ascending path, one that could take thousands or millions of steps to traverse. To find out if such a path exists, Kauffman and Levin first computed the average number of steps that it would take a population, starting out at some arbitrary place, to ascend the nearest peak, from which natural selection could go no further. They found that the number of steps to the nearest peak was paltry, smaller than fifteen, and not nearly enough to get a population to Mount Everest.[10] Most populations would wind up on the nearest molehill.

Theoretical calculations like this cannot replace experiments that sketch the true contours of a fitness landscape, count its peaks, and trace all access routes to them. Unfortunately, such experiments will never be able to map any one landscape completely, because genomes have so many variants, but they can focus on smaller regions, such as those where only one gene varies. This is useful because any one gene encodes a protein, and proteins are not only the workhorses of our cells, but also crucial links between the genotype and the phenotype. Each cell harbors thousands of different kinds of proteins, each encoded by a gene, each with a different task. Proteins are among the smallest parts of an organism with a phenotype worth studying.

Each protein is encoded by a string of DNA letters, and the collection of all possible such strings—also called a space of sequences—is a giant realm of possibility. Think of it as a library of texts that encodes not only all of the countless innovative proteins that evolution has discovered in its history, but also all the proteins that it could discover in the future. It is the space where nature goes to find new parts for its biochemical machines.[11]

To map the fitness landscape of this space is to measure how well each of its DNA sequences—or, equivalently, the amino acid sequences they encode—is suited for a specific task. It is to measure the speed at which protein enzymes can cleave a sugar molecule, the pull motor proteins can exert in a muscle, or the rate at which transport proteins can ferry nutrients into a cell. And because this landscape's topography channels the movement of populations—uphill, always uphill—it also imposes limits on nature's creativity in finding new and better proteins.

Unfortunately, even this library of protein texts is too large to explore completely—there are more than 10^{130} proteins with one hundred amino acids each, and many proteins are much longer. Therefore, experimenters must focus either on shorter strings, of which there are fewer, or on a smattering of paths through the landscape. Even that requires technologies to manufacture numerous DNA and protein strings, and it had to wait until the first decade of the twenty-first century, when such technologies became efficient enough, eighty years after Wright birthed the landscape idea.

Some such paths through the vast library of protein texts lead to innovations that save lives—not those of humans, but of our lethal enemies, disease-causing bacteria. These bacteria have discovered proteins called beta-lactamases that disarm the offensive weapons—antibiotics—that doctors use to kill bacteria. Named after beta-lactam, a ring of atoms that occurs in antibiotics like penicillin, beta-lactamases can destroy this ring and defuse these antibiotics. Because beta-lactamases save bacteria from death, they spread like wildfire through bacterial populations—courtesy of natural selection—while endangering the lives of patients, who are left helpless when overrun by a bacterial infection. Innovations like beta-lactamases are nature's defenses in an endless arms race between medical researchers, who constantly develop new offensive weapons, and bacteria, whose vast populations scour nature's DNA libraries for new ways to neutralize these weapons.

An especially important offensive weapon is cefotaxime, a broad-spectrum antibiotic capable of destroying many different kinds of bacteria. It is on the World Health Organization's List of Essential Medicines. Alas, it may not remain there much longer, because of a simple disturbing fact: disarming cefotaxime requires nothing more than a few tweaks in today's beta-lactamase proteins.

A conventional beta-lactamase disarms cefotaxime slowly, too slowly to help bacteria survive the hefty doses prescribed by doctors. But it turns out that with only five letter changes in such a beta-lactamase the protein is rendered one hundred thousand times more efficient at destroying cefotaxime.[12] The new protein variant is a high peak—although perhaps not the highest—in the fitness landscape of proteins that can destroy cefotaxime. How hard is it to reach this peak? Is it the top of a craggy mountain or a smooth sugar cone? The ideal experiment to answer these questions would manufacture all protein variants around this peak, measure their ability to destroy cefotaxime, and find out which of them are lower outcrops that could stop selection's march. Alas, their number is too large. There are more than a trillion proteins that differ from conventional beta-lactamase by five or fewer amino acids, more than can be easily manufactured with current technologies. But even though we cannot map every local peak and valley, we can still get a glimpse of the whole by following individual paths.[13] Here is how.

Imagine you are blind, standing at the foot of a mountain and wanting to climb it. You cannot see the best path to the peak, but you can distinguish an uphill from a downhill step, so you feel your way as you take step after step. If the mountain is perfectly smooth, then every path of uphill steps will eventually lead to the peak. Some of these paths will meander uphill in serpentines, others might spiral slowly toward the peak, while yet others will lead straight up, but every single one will eventually get you there. Not so if that mountain is craggy. In that case, most paths will get you stuck on some outcrop below the peak, and only a few—not necessarily the straight uphill ones—may lead all the way up. To find out how craggy the mountain is, you can do this: try the same climb multiple times, with sequences of steps in different directions, but all uphill, and count how often you get stuck. If all your attempts succeed in reaching the peak, the mountain is perfectly smooth. If you get stuck every time, it is maximally rugged.

In 2006, Daniel Weinreich, then a postdoctoral researcher at Harvard University, translated this very idea into an experiment on beta-lactamases, tracing paths from the original beta-lactamase protein to its cefotaxime-destroying variant in which five amino acid letters are changed. Each letter change is one step toward the peak. Because five letters can change in different orders, different paths lead up the peak, just like you can transform the word BOLT into GOLD by editing two letters in different ways, either from BOLT to MOLD to GOLD, or from BOLT to (the meaningless) GOLT to GOLD. There are 120 different orders in which five different amino acid changes can occur, each of them a different path toward the cefotaxime peak. Weinreich and his collaborators synthesized all proteins along each path and measured their ability to destroy cefotaxime in order to identify which paths are dead ends.

Most of them are, it turns out. More than 90 percent of the paths lead some way up the peak but then encounter a protein that cannot be improved by a single further step. And because natural selection prohibits backtracking downhill, evolution's uphill climb would end right there.[14]

A dozen further experiments run by other researchers like this have climbed peaks elsewhere in the vast landscapes where molecules and organisms evolve. They created bacteria that can grow and divide faster on the same diet, HIV viruses that can infect human cells more efficiently, and enzymes that manufacture new self-defense chemicals for plants. And they reveal a similar topography, one not as hopelessly rugged as the landscapes studied by theorists like Kauffman and Levin, but still much more rugged than a sugar cone.[15] Among the many paths that lead toward a peak, only a few reach it. On the remaining paths, natural selection dead-ends below the highest peak—sometimes far below. In evolution's landscapes, a climber's risk of getting stuck near base camp is very real.

———

For billions of years, proteins have produced a steady stream of innovations, but another kind of innovative molecule—RNA—has been at it even longer. This ugly duckling of molecular biology, long thought to be a mere carbon copy of DNA, helping to make protein, metamorphosed into a swan when biochemists discovered in the 1980s that it can do so much more.

Like protein, RNA can catalyze chemical reactions, but unlike protein, its letter sequence also stores the same kind of heritable information as DNA. Its special talents have helped cast RNA as the leading actor in some of the invisible dramas that play out inside every living cell. For instance, RNA collaborates with proteins in an enzyme called telomerase by helping to maintain pesky chromosome ends called telomeres, which tend to shorten over time, a bit like fraying shoelaces, except their fraying has more serious consequences. When it goes unchecked—for example, because the telomere maintenance crew is too slow—cells quickly stop to divide, age, and die. That's bad, but even worse is when that telomerase is hyperactive, because then cells can start to divide uncontrollably and become cancerous.

Another RNA-equipped biochemical machine is just as remarkable because it opens a window into the very early history of life itself. This is the protein-manufacturing ribosome, a hugely complex apparatus comprised of several RNA strings and more than fifty proteins. Among all these molecules, RNA plays the most important role because it is one of the ribosome's RNA molecules—transcribed from a special RNA-encoding gene—that performs the crucial task: stringing amino acid parts together, letter by letter by letter, to build a protein string. The ribosome is one of several clues that early life was an RNA world, one where RNA ran the show the way proteins do now.

Another telltale remnant of this sunken empire is a bizarre process that allows some genes to encode multiple proteins. Once a gene's DNA has been transcribed into an RNA carbon copy, a cell sometimes deletes short pieces of that copy and splices the remaining pieces together. When the same gene is transcribed twice or more, these deletions can occur in different places, creating different RNA transcripts that are translated into different proteins. Alternative splicing, as biochemists call it, is a nifty mechanism to create proteins with different functions from the same gene. It's as if you created many shorter variants of a long poem by combining a few lines here and there, using a different combination each time. In human language, most such variants would be garbled and nonsensical, but in the chemical language of proteins, they can encode meaningful, useful proteins. And even though alternative splicing may seem bizarre, it can be quite important. For example, it produces variants of a human protein required to detect sound. These variants help tune cells in our inner ear to perceive sounds of different frequencies.[16] No alternative splicing, no Bach, Bartok, or Beethoven.

For this kind of creative editing, complex organisms like humans need another complex biochemical machine called the spliceosome, but simpler ones like bacteria don't.[17] What's more, in some bacterial genes the transcribed RNA string *itself*—without any help from proteins—can do the job. It discards part of its own text and splices whatever is left into a new, shorter string. Such a wonder molecule is not only an RNA enzyme—biochemists call it a ribozyme. It is an RNA enzyme that modifies itself. Think of it as a poem capable of rearranging itself on the written page.

Just like proteins, RNA molecules are texts written in a molecular alphabet—four nucleotide letters instead of proteins' twenty amino acid letters—that form a library vast beyond imagination. Some of this library's texts are capable of self-splicing, and one of them occurs in the genome of an otherwise unremarkable soil bacterium called *Azoarcus*. Eric Hayden, a young researcher in my laboratory at the University of Zurich, used the *Azoarcus* ribozyme as a base camp to climb a nearby peak in its fitness landscape.[18]

Eric knew that this RNA molecule could use its self-splicing powers to join itself to another string of RNA with a specific letter sequence, while it would fail miserably at joining itself to a third string with a different letter sequence. But in earlier experiments, Eric had found a more flexible ribozyme that could self-splice with both strings. This ribozyme was at the peak he wanted his molecules to climb, a peak that was only four letter changes away from the *Azoarcus* ribozyme. Eric synthesized all RNA molecules that lie between the *Azoarcus* ribozyme and this peak, which allowed him to study every single one of the twenty-four possible pathways that lead up the peak. He discovered that only one pathway actually leads all the way up—the others are impassable for natural selection, because they lead through valleys in the landscape. These experiments taught us that RNA fitness landscapes can be just as rugged as those of proteins.

Researchers like Eric explore a fitness landscape the hard way, painstakingly synthesizing the molecules on all different paths to the peak. (His molecular journeys took more than a year of dedicated laboratory work.) Other researchers use automated synthesis technology to manufacture huge collections of molecules. This technology can enable them to catalog an entire library of molecules. The drawback is that this works only for small libraries comprised of molecules much shorter than the two-hundred-plus letters of beta-lactamase and the *Azoarcus* ribozyme.

Actually, small is relative. In one such study, researchers at Harvard University created all possible RNA molecules that are twenty-four nucleotide letters long—more than 280 trillion of them—and scoured this library for molecules with a skill that is essential to all life: they can attach themselves to other, energy-rich molecules.

This skill is essential because many chemical reactions inside an organism require energy. Much of this energy comes from molecules that store energy in the chemical bonds linking their atoms. But to use it, an enzyme—protein or RNA—must first attach to an energy-rich molecule so that it can harvest this energy. Hence the question these researchers asked: Which among these trillions of short RNA molecules can perform this first step of energy harvesting?

In their experiments, the researchers used an energy-rich molecule called guanosine triphosphate, or GTP, that is recognized by molecules in every living organism, and they found not just a few but thousands of such GTP-binding RNA molecules.[19] What's important is that these RNA molecules did not form a single peak in the fitness landscape of energy-harvesting molecules. Instead, they clustered into fifteen different fitness peaks that differed in height—the higher the peak, the greater an RNA's attraction to GTP. What is more, the peaks were scattered far and wide throughout the landscape. A population of molecules stuck at a low peak could be stuck there forever.

Not all creations of biological evolution rely on new molecules like the energy-hungry RNAs or the antibiotics-devouring beta-lactamases that I described. Some require nothing but changes in the place and time where an old molecule is made. That's because the unfolding of new life in a developing embryo follows a program, a recipe a bit like that in a cookbook, but sophisticated beyond belief, requiring thousands of protein ingredients and exquisite timing when adding these ingredients to the simmering stew. By changing only the when and where of adding these ingredients, evolution can bring forth entire new life-forms—four-legged animals from fish, feathered birds from dinosaurs, and so on. Here is how.

Our bodies harbor trillions of cells, but only a few hundred different *kinds* of cells, like those that transmit electrical signals in your brain, contract muscles in your arms, or transport oxygen in your blood. Each cell type harbors different molecules—many of them proteins—that are unique to it, much like a fingerprint is unique to a person. In other words, each cell transcribes and translates into protein only some of the twenty thousand genes of our genome. Some genes are only turned on in the liver, others only in the brain, yet others only in muscle, and so on. Where, when, and how often genes are transcribed and translated is regulated by specialized proteins known as transcriptional regulators. These cooperate with the biochemical machinery that transcribes a gene into RNA. The details of this cooperation are complex, but the basic principle is simple: to make their influence felt, these regulators need to be close to where the biochemical machinery starts transcribing, which is at the gene's beginning. They achieve this with a very simple mechanism. Each such regulator protein can recognize and latch onto short DNA words with specific sequences of letters, like CATGTGTA or AGCCGGCT, and if such a word occurs near a gene, the gene's transcription gets turned up or down. What is more, many of the thousands of genes in our genome contain DNA words recognized by the same regulator. In this way, one regulator can regulate a multitude of genes.

When our bodies grow and develop from a fertilized egg, these regulators—hundreds of them—are like cooks that follow the immensely complex recipe needed to build a body. They ensure that thousands of genes are turned on to just the right level, helping to manufacture the right proteins in the right amount. Nothing illustrates the importance of these regulators better than the birth defects that happen when they fail to follow this recipe to the letter. A minor glitch will create a mild birth defect like a cleft lip or fused fingers, while major deviations will lead to serious defects like malformed hearts or even to lethal ones where a mangled body dies before it is born.

Because gene regulation helps sculpt all multicellular organisms, from primitive jellyfish to complex primates, from microscopic algae to gargantuan redwood trees, a new kind of body—or even just a new body part—requires new regulation. The tubular bodies of snakes are braced by a grotesquely elongated thorax containing hundreds of ribs, the long legs of horses are supported by a massively enlarged third toe that helps them outrun predators, and in some orchids a simple whorl of petals is transformed into an elaborate lure whose resemblance to a female insect is uncanny, attracting male pollinators on the prowl for females. All these products of life's creativity require altered regulatory recipes that guide genes to making a bit more or less of their proteins a bit earlier or later, subtle alterations that manipulate the numerous ingredients needed to create new life.

Evolution can manipulate such recipes easily because any one regulator recognizes not just one but hundreds of different DNA words. It binds to some of them tightly, stays on the DNA for a long time, and turns a nearby gene on strongly, like a volume dial cranked up to the max. Others it binds to loosely, falling off soon after it has latched on, such that the gene is barely turned on, the volume barely audible. Single-letter changes in one of these words can modulate a regulator's binding and fine-tune how often a gene is transcribed. Many such small changes can add up to a new body architecture, a new creative product of evolution.

Together, all these DNA words—each a special kind of genotype—form a landscape of gene regulation. But unlike some of the hopelessly vast landscapes we have encountered, we can map this landscape completely because the DNA words bound by most regulators are short, usually fewer than a dozen letters. For example, compared to the astronomical number of proteins with one hundred amino acids—more than 10^{100} texts—there are only sixteen million twelve-letter DNA words.

Such smaller numbers are a relief for those of us who study molecular landscapes because we have a technology that allows us to measure how strongly a regulator binds to every single DNA molecule in such a landscape. It is known as microarray technology or DNA chip technology. Like the chips in a computer, which perform many simple calculations simultaneously, a microarray allows scientists to perform many measurements in one go. Think of a microarray as a rectangular grid with as many locations—grid points—as there are DNA words to be studied. Each location harbors many copies of a DNA molecule with one specific letter sequence. When bathing the chip in a solution containing a regulator protein, the regulator will bind to some DNA words, and the strength of this binding can be measured on each grid point.[20] In short, a single DNA chip experiment can map an entire fitness landscape. If an orchid's flower is maximally seductive, if a fruit fly's wing creates maximum lift, or if a horse's leg provides optimal support only if some regulator turns a specific gene on to the max, then the peaks of this landscape are the regulator's most tightly bound words. And because microarray technology makes it so easy to map an entire landscape, it has been used to map the landscapes of not just one or a few, but more than a thousand regulators, from organisms as different as plants, fungi, and mice.[21]

Such microarray data allowed Joshua Payne and José Aguilar Rodríguez, two researchers in my Zurich laboratory, to ask a question that will sound familiar by now: How many peaks do these landscapes have?[22]

The answer mirrors what we learned from other adaptive landscapes: the landscapes of gene regulation can be somewhat rugged but not impossibly so. Many of them even have only a single peak—easy to conquer by natural selection—but others have dozens of peaks with different heights. The peaks correspond to different DNA words that are able to turn on genes to varying extents—some more, some less—but from each peak, higher peaks cannot be reached by walking only uphill. The landscapes on which evolution can explore new body architectures are no different from those that brought forth its other creative products, including proteins that disarm new antibiotics and RNA enzymes that can splice new RNA strings.

Biology has come a long way since Sewall Wright's day, when he could only speculate on the topography of evolution's landscapes. His hazy guesswork has been supplanted by high-resolution maps containing the finest molecular details, like satellite imagery that can resolve single grains of sand. But even more important than these details is an idea that is central to any science of creation. Wright was unaware of its generality, but we will encounter it again and again in later chapters: the difficulty of a problem can be encapsulated in the topography of its landscape. Single-peaked smooth landscapes correspond to easy problems. Their peaks—harboring the single best solution—can be conquered through exclusively uphill steps. Multipeaked landscapes correspond to harder problems. The more peaks, the harder the problem. The hardest problems need the most creative solutions, and a big part of finding these solutions is getting off of dead-end peaks and finding higher ones.

Every example of a fitness landscape so far in this book has revolved around a problem that nature solved—from efficient swimming by bulky ammonites to disarming novel antibiotics unleashed on bacteria. For easy problems, a straight uphill march will do, but unlike von Helmholtz, who could retrace his steps from a dead-end peak, populations driven uphill by natural selection cannot. This is an important lesson for those who call natural selection all powerful: selection's relentless uphill drive—like that of a hypercompetitive human who always strives for the faster, better, superior—is a fatal impediment to solving truly difficult problems. Selection and competition alone are impotent to solve such problems.

Which leaves us with a crucial question: How does nature get off those dead-end peaks?

Chapter 3

On the Importance of Going Through Hell

In 1922 British archeologist Howard Carter discovered the tomb of King Tutankhamun, where he found 130 walking canes, even though King Tut died at only nineteen years old. Somebody had clearly thought that King Tut could use these canes in the afterlife, but it would take almost a century to find out why. That's when a research team led by the Egyptian archeologist Zahi Hawass discovered through a CT scan that King Tut had suffered from various deformities, including a left club foot and a missing toe on the right foot, as well as a cleft palate and signs of a hereditary bone disease. Tut could really have used those canes. Historical records and DNA testing show that incestuous marriages were rampant in the bloodline leading up to King Tut. Tut's parents, for example, were brother and sister. Further DNA work also resolved another mystery: the identity of two stillborn fetuses interred in King Tut's tomb. They turned out to be Tut's children—failed attempts to continue the royal bloodline.[1]

Perhaps King Tut's was the first royal bloodline that succumbed to the dangers of inbreeding, but it certainly was not the last. Three thousand years later, a similar fate befell the European Habsburg dynasty.

The Habsburgs were an ugly bunch. Walk into any major European museum of art, and you will recognize their portraits without reading a label. What gives them away is that oddly jutting lower lip, well known as the Habsburg lip.

Their misfortune, it turns out, was more than cosmetic, an ominous sign of much deeper maladies. The Habsburg lip resulted from mandibular prognathism, a protruding lower jaw so large that the lower and upper teeth no longer align. This deformity itself emerged from six centuries of incestuous intermarriage among a small circle of royal families. Their marriages helped forge political alliances, prevent wars, and gain new territories. Long before twentieth-century geneticists would explain why inbreeding has disastrous consequences, these and other aristocratic families experienced its effects firsthand. In the Spanish Habsburg line alone, nine of eleven marriages, from King Philip I in the fifteenth century to Charles II in the seventeenth, were consummated between cousins or between a man and his niece. Even an observer completely ignorant of genetics would have known that something was amiss: childhood mortalities in the royal family were as high as 50 percent, more than twice the level found among ordinary Spaniards.[2]

At the end of this multigenerational experiment stood King Charles II, who had much more wrong with him than just his lower lip. He was barely a notch above a drooling idiot. Unable to speak until age four, unable to walk until age eight, he was short, weak, and thin, with a tongue so large his speech was difficult to understand and a lower jaw that protruded so far that he could not chew. He was uninterested in the world around him and, worst of all, was plagued by a disease fatal for any royal bloodline: impotence. And so with him the Spanish Habsburg line went extinct.[3]

Deformities of body and mind are among the multiple consequences of inbreeding, but don't be misled into thinking that inbreeding is all terrible. It can also be a force for good. Animal breeders, for example, rely on it to enhance coveted features of livestock or pets. That's because inbreeding—unlike natural selection—is indifferent to whether a feature is good or bad. It merely brings out extreme features. A cattle breeder may breed a single Texas Longhorn bull that has exceptionally long horns with multiple females. All of their offspring are half-brothers or half-sisters, and some of them may have horns just as long as their dad. The breeder might mate those exceptional animals with each other and cull the rest. Continue this selective breeding for several generations, and the exceptional horns of the founding father can become ordinary in his distant descendants.

Selective inbreeding like this—among livestock, pets, or plants—is a time-honored procedure. But as the Habsburgs and King Tut show, it also has unintended consequences. These consequences can be understood from basic genetics. Let me explain.

The genome of a bull has two copies each of its more than twenty thousand genes, one from its mother, the other from its father. Because mutations constantly sprinkle any genome with DNA errors, the DNA sequences of the two copies will differ for many of these genes, and in some genes a mutation may have damaged one of the copies. That's usually no problem, as long as the other, intact copy is around. But when two damaged copies come together in the same genome, the result is a genetic disease.[4] This is usually rare. It can happen, for example, if the bull had already inherited a damaged copy from its parents and the other copy suffers a mutation during the bull's lifetime. But it becomes frequent when the descendants of the same family breed repeatedly. Each of the bull's offspring—all of them half-siblings—has a 50 percent chance of inheriting a damaged copy of the same gene from its father, and if these half-siblings themselves sire offspring with each other, the chances that their offspring get the disease are large—25 percent, to be precise, as geneticists can calculate. Breeders use various tricks to reduce the number of diseased animals, such as periodically outbreeding some individuals or culling the sickest animals, but they cannot completely avoid disease.

Because there are so many genes that could be damaged in any one family line, inbreeding brings out different defects in different lineages. Some of them may be mild, such as a missing tail switch in cattle. (It is considered an aesthetic flaw by cattle breeders, but to a cow it surely would be a handicap in keeping those pesky blood-sucking insects away.) Others are less innocuous, like shorter fur in cattle. It means poor thermal insulation in the winter, resulting in calves that gain weight more slowly—not exactly a desirable trait in beef cattle. Yet other defects are disastrous. They include premature death and low fertility, and neither of these is rare, just like in those royals.

Because inbreeding indiscriminately affects both good and bad traits, various races of cattle, dogs, cats, and other domestic animals are defined as much by their flaws—often hidden—as by their strengths.

German shepherds frequently suffer from a malformed hip joint, whose ball does not fit tightly into the joint's socket. As a result, many of these beautiful animals walk in pain, labor to climb stairs, strain to get up, or become lame as they age.

Even worse off are American Burmese cats. They are a gorgeous breed of cat with wide-set and large, vaguely childlike and curious eyes. Sadly, some of them also suffer from a horrific and lethal deformity called the Burmese head fault, where some kittens develop two upper muzzles, and the top of the head is incompletely formed. Their misfortune can be traced to a single stud cat that sired many prizewinning offspring, but also introduced the deformity to the breed. The stud's name was Good Fortune Fortunatas. It surely didn't bring any good fortune to those pitiful kittens.

In some animals, the very gene that causes a coveted trait also brings about its opposite. A case in point is Ojos Azules, a stunning Mexican breed of cats with—the name says it all—deep blue eyes. The blue color, really a lack of pigment in the iris, results from a mutation in one gene, and as long as only one of the gene's two copies is mutated, everything is fine: eyes are blue and kittens are healthy. But woe unto the cat whose copies are both mutated. It suffers from a deformed skull and will not even survive birth.[5]

Nature has developed mechanisms to avoid inbreeding wherever it can. In some animals, offspring scatter far and wide, as in prides of lions, where all males and some females leave the natal pride and disperse to find their reproductive luck elsewhere.[6] In other animals—mice, quail, and voles—familiarity must breed contempt, or at least lackluster sexual attraction, because littermates or nest mates rarely mate.[7] In humans, this phenomenon is called the Westermarck effect, named after a Finnish anthropologist who first proposed that growing up together suppresses sexual attraction among adults. Case in point: kibbutzim, the Israeli communities that raised children collectively, where very few marriages occurred among the thousands of men and women raised in the same kibbutz.[8] Inbreeding is even bad for plants, where it is a prelude to stunted plants, seeds that do not germinate, and "albino" seedlings unable to harvest light.[9] The flowers of some plants detect the molecular fingerprint of an incoming pollen grain, and, if this fingerprint is too similar to their own—the grain comes from a close relative or even from the same plant—the pollen grain fails to fertilize, and no seed develops. To be sure, some of these behaviors may also exist for other reasons. Roaming lions, for example, also avoid sexual competitors. But all of these behaviors help avoid inbreeding.[10]

Unfortunately, try as they might, nature's children are sometimes forced into incestuous relationships. Whenever a population is decimated by a cataclysmic storm, hunted to near-extinction, or swept onto a small and remote island, only a few individuals may be left to keep the population alive and the gene pool from drying up. Among all such disasters that can befall a population, exile on a small island is especially fascinating to biologists. That's not only because sandy beaches and unspoiled forests make great destinations for field research. Islands are like isolated laboratories where one can observe evolution in action, and this action holds a key to understanding the creativity of nature—and of much more than nature, as we shall see in later chapters.

Imagine four unrelated people, two men and two women, who are shipwrecked on a tiny Pacific atoll that does not provide space, water, or nourishment for many more. Not having much choice in the dating game, they form two couples and have children. The children of the two couples are genetically unrelated, and if each of these children, once grown, has children only with a member of the other couple, *their* children—the second generation—will also be unrelated. But when these grandchildren are ready to have children themselves, inbreeding will become unavoidable. The simple reason: all members of the second generation share the same grandparents, or about one quarter of their genes. From that generation onward, all individuals on the island will be part of the same family.

The same thing happens—everybody eventually becomes genetically related—even if the starting population is a bit larger. It just takes longer. As a rule of thumb, it takes some ten generations for ten initial individuals, or one hundred generations for one hundred individuals, to congeal into a single big, but perhaps not genetically happy, family. The more individuals, the longer it takes, but eventually a population of any size will have the same fate—everybody will be everybody else's relative, although some individuals will be more inbred than others.[11]

Whether enforced by circumstances or by a breeder, inbreeding has the same consequence: genetic trouble. Manx cats from the Isle of Man, named after the Celtic natives of this tiny, 570-square-kilometer island in the Irish Sea, could tell you a story or two about this trouble. They do not have a club foot like King Tut, but their skeleton has a defect no less obvious, born from multiple generations of inbreeding: it lacks a tail. That defect impairs a cat's balance and precludes some of the acrobatic feats that cats are known for. But it is not the worst problem of this race. Many kittens have deformed spines, or are even stillborn, and litter sizes are generally small—low fecundity. Fortunately, these disadvantages are offset by some advantages, among them the great hunting skills that made Manx cats popular as mousers on farms, and a playful, almost doglike disposition cherished by some owners.

The rough lives of inbred Soay sheep on the island of Hirta tell a similar story. Hirta is part of the windswept, treeless, and rugged St. Kilda archipelago off the Scottish west coast. The archipelago has been uninhabited since its last thirty-six human inhabitants evacuated to the mainland in 1930 once they'd had enough of their harsh life. The Soay sheep—*Soay* means *island of sheep* in old Norse—lacked this option, and maybe they did not mind, because they have lived on the islands in small numbers since time immemorial. Yet their life is no bed of roses, because many of them are plagued by intestinal worms that siphon off the scarce nutrients the sheep ingest. And when the most inbred sheep die during the winter—up to 70 percent of the whole population perishes in bad times—it is really the overwhelming number of worms that kills them, because inbreeding has weakened their immune systems.[12]

The same pattern can be found on Mandarte Island, a truly tiny Canadian island of almost six hectares off the coast of British Columbia, home to some one hundred song sparrows whose numbers fluctuate over the years. When a severe winter storm in 1989 left only eleven surviving birds, it had killed off the most inbred ones.[13]

Club-footed King Tut, worm-plagued sheep, and numerous other cases of inbreeding underscore that natural selection is not all-powerful; its relentless uphill pull in nature's adaptive landscapes can be stymied. To understand when and why that happens, we need to understand a phenomenon that has affected life since its origins, even though it has only been appreciated since the early twentieth century. That's when biologists like Sewall Wright first saw that studying entire populations and not just individuals is crucial to understanding evolution. Today, the phenomenon is called *genetic drift*, but for some time it was called the Sewall Wright effect, in honor of Wright's contributions to its discovery.[14]

Wright was among the first to view a population not just as a collection of organisms but as a pool of genes or alleles. That shift in perspective is key to understanding genetic drift. Small populations have a small gene pool, and large populations have a large one. Let's imagine a *very* small population with a pool of only four alleles that influence some trait, such as a person's eye color. Imagine that our small gene pool contains two alleles for brown eyes and two for blue eyes. To understand how genetic drift affects the fate of these alleles over many generations, it is useful to think of this gene pool as a bowl containing four marbles of two different colors—such as the black and gray circles on the left-hand side of Figure 3.1.[15]

In the early twentieth century, geneticists already realized that passing on genes from one generation to the next—creating the next generation's gene pool—works just like a series of blind draws from such a bowl. So, to create a new gene pool, let's first blindly draw a marble and note its color—brown or blue—put it back in the bowl, then draw another marble (still without looking), note its color, put it back, and repeat the drawing and replacing a third and fourth time to create a new pool of four genes. (Putting the marble back before drawing again captures an important feature of inheritance: different sperm or egg cells can inherit a copy of the same allele from one of their parents.)

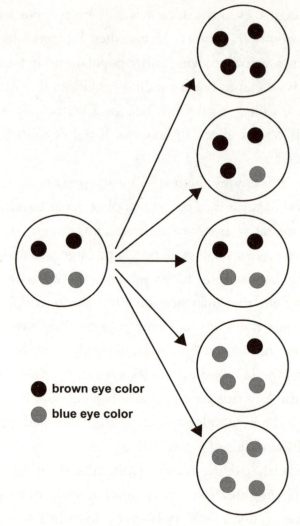

Figure 3.1.

At the end of this repeated drawing, we have picked four
marbles. If you tried this at home, you would see that these marbles will not necessarily show the same color combination—two
brown and two blue—as the parental gene pool. Sometimes we
may get three brown and one blue marble, four brown ones, or,
conversely, three blue and one brown marble, or four blue ones
(Figure 3.1). The number of marbles of a given color corresponds

to the frequency of the alleles of a different type. The main point is this: their number varies randomly from one generation to the next.

To see how such a gene pool would evolve over multiple generations, we can draw four marbles from the new pool in the exact same way to form a third-generation gene pool, then repeat that drawing to generate a fourth generation, and so on, ad infinitum.

Early twentieth-century population geneticists used the mathematics of probability theory to find out how this gene pool will change in the long run. The math is complex, but its main lessons are simple. First, the number of marbles of any one color—alleles of any one type—continues to change randomly and unpredictably until alleles of only one type—brown or blue—are left. When that happens, one of the two alleles has become extinct, whereas the other, in population genetic jargon, has become *fixed* in the population. The gene pool will remain unchanged from that point onward unless a DNA mutation recreates the other allele, and the chances of that are minute.[16]

The math also shows that while one allele will eventually become fixed—with certainty—which allele will be the lucky one is as unpredictable as the outcome of a coin toss. In 50 percent of populations, the blue allele will become fixed, and all individuals will end up having blue eyes. In the other 50 percent, all individuals will have brown eyes.

Even though human genes get reshuffled in complicated ways when our bodies make the sperm and egg cells needed to have children, the drawing of marbles from a bowl captures exactly how genes and their alleles are passed from generation to generation.[17] And that we humans have two copies of each gene makes no difference either: a pool of four human genes corresponds to a "population" of two people, each with two copies of each gene. What I just described is how this gene pool would evolve if these two people had two children, a girl and a boy, who grow up and have another girl and a boy, who have another girl and a boy, and so on. Eventually, this highly inbred family would have all blue or brown eyes, and nobody could tell in advance which it was going to be.[18]

Genetic drift affects larger populations as well. As a rule of thumb, in a pool of ten genes—five individuals with two alleles each—allele numbers fluctuate by about 10 percent each generation. In a pool of one hundred genes, they fluctuate by only 1 percent, and in a pool of a thousand, by 0.1 percent. Moreover, the time a blue or brown allele takes to spread through a gene pool and become fixed is longer in large populations. For a gene pool that is ten times as large, it would take ten times as long for all individuals to carry the same allele. That's why inbreeding is not as much of a worry in very large populations. Yes, drift does affect a gene pool of a billion, but it causes only tiny jitters in allele numbers, causing them to fluctuate by 0.0000001 percent every generation.[19]

A DNA mutation creates only a single new copy of any one allele, and only in a single individual of a population. For the many genetic diseases where two bad copies need to come together in the same individual, this rare allele can spread through a population via genetic drift, generation by generation, unmolested by natural selection, until the allele has become frequent enough that two bad copies begin to come together in some individuals. When these individuals die as a result, natural selection is kicking in. The smaller the population, the faster this will happen. And because we have so many genes, chances are high that bad alleles will come together for at least one or a few genes.[20]

That's where the diseases caused by inbreeding come from, and that's why different diseases occur in different inbred populations. If you could reset evolution's clock at will and start to inbreed the same small population, you would find that the bad alleles that genetic drift helps spread are different each time.

All this means that the negative effects of inbreeding are a consequence of genetic drift in small populations. It does not matter if an island forces a population to be small, if royal politics—or conceit—restricts an aristocrat's potential mates, if a mountain range isolates a population of villagers, or if a breeder repeatedly mates individuals from the same lineage. The result is the same: genes that would otherwise be wiped out can persist and spread through a population.

Once again, this doesn't mean that drift and inbreeding necessarily have bad consequences. Both are simply indifferent to the concept of good or bad—they will spread bad alleles just as they spread good ones. What's more, genetic drift is not necessarily going to lead to inbreeding. Some single-celled fungi and algae have only one copy of each gene, which means that the coming together of two bad alleles is impossible. Bacteria also have only one copy of each gene, and they reproduce without sex as we know it, so they, too, are not subject to inbreeding.[21] Yet genetic drift is at work in their populations, just as it is at work in those of algae, fungi, or any other organism, because the blind drawing of genes to fill a gene pool is fundamental to all life-forms, regardless of how they reproduce. Inbreeding is not universal to life, but genetic drift is.

To visualize what genetic drift does to a population, recall the fitness landscapes of the previous chapter, and the relentless uphill pull that natural selection exerts on populations evolving on these landscapes. Because drift causes random and direction-less changes in a population's gene pool, you can think of it as an unceasing tremor, like an earthquake that causes this land-scape to tremble and shake without pause.[22] Any mountaineer's ascent would be slowed by an earthquake, and a hill-climbing population is no exception. Because genetic drift is direction-less, these tremors can take the population in any direction—uphill, downhill, or sideways (Figure 3.2). If the tremors are weak—in a large population—they will not delay the climbing population by much, but the stronger they are—the smaller the population—the more violently they will jerk the population around, and the more they will slow it down. Downhill lurches

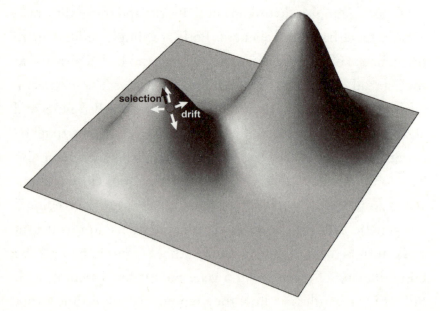

Figure 3.2.

are especially momentous because they can bring a population to a valley where it does as poorly as those inbred royals and worm-infested sheep.

The genetic math can also tell us how strong the tremors—how small the population—must be to override the uphill pull of selection. Take a population whose organisms differ by 5 percent in their fitness—one type of bacterium dividing 5 percent faster than another, one apple tree producing 5 percent more seeds, or one squirrel with a 5 percent greater chance of surviving a harsh winter. In such populations, drift can overpower selection if it jiggles allele numbers by more than 5 percent. That happens in populations of twenty or fewer individuals.[23] If, however, a population's individuals differed in their fitness by only 1 percent, drift could overpower selection even in somewhat larger populations because it has to jiggle allele numbers by much less—only more than about 1 percent—to overcome selection. The math shows that in this case, populations with fewer than one hundred individuals are small enough to create jitters strong enough to overpower selection and prevent the population's ascent. The general pattern is simple: if individuals differ in their fitness by a tenth of a percent, drift could overpower selection in a population smaller than a thousand individuals, and if they differ in fitness by a hundredth of a percent, a population of fewer than ten thousand individuals would suffice.

Fitness differences as small as these are important in evolution because they are natural selection's most abundant raw material. We know this because we can measure how strongly mutations alter fitness. The vast majority turn out to change fitness just a wee bit.[24] Such small effects, however, can accumulate since evolution unfolds over millions of generations or years.

In the long run, even mutations that cause fitness differences smaller than a millionth of a percent can mean the difference between survival and eventual extinction[25]—as long as they occur in a population larger than 100 million individuals. In a population that is any smaller, selection would be powerless to overcome genetic drift.

I have used alleles that affect eye color to illustrate the action of random genetic drift because eye color seems an especially innocuous trait, but it turns out that people with lighter, bluer irises are slightly more prone to getting some cancers of the eye.[26] Thousands of other genes in our genomes affect traits much less innocuous than eye color—the strength of our bones, the vigor of our immune system, or plain old fertility. Because all genes are inherited according to the same rules, "bad" alleles of these genes can also spread through a population via random genetic drift as long as a population is small enough and drift is strong enough.

In the fourteenth century, the poet Dante Alighieri reached a summit of world literature with the epic known as the *Divine Comedy*. The creative process guiding him is lost to history, but for some creators, this process may resemble the journey that the *Divine Comedy*'s protagonist has to endure: he must descend through the nine circles of hell, whose inmates suffer increasingly gruesome and imaginative forms of torture, before he can eventually ascend through purgatory into the nine celestial spheres of heaven.

Those philistines who are indifferent to this epic's flowering allegories might just see it as an elaboration of an age-old principle: for things to get better, sometimes they first need to get worse. This principle, it turns out, is much more profound than they realize because it applies far beyond the human realm. Life's four-billion-year-long journey of evolution would not have gotten very far without it. That's because a population stuck at a dead-end, low peak of an adaptive landscape cannot be led away from it by natural selection, which is powerless to go anywhere but up. Genetic drift, however, can help the population descend into the hellish cauldron where new and successful combinations of genes are forged.

Unfortunately, descending into this genetic hell is more than just dangerous: many populations never even reach purgatory. They simply go extinct. And only their remains can tell us about their tragic fate.

These remains are especially eloquent on islands, where colonizers are few and genetic drift is strong. On Hawaii alone, more than 30 percent of all flowering plant species that colonized the islands never made it. And insects fared even worse. A combination of ill-adapted genes and a hostile environment extinguished more than 150 of their species, some 80 percent of the original colonists.[27]

But to those populations that were lucky enough to persist and turn the vale of tears into their base camp, many new peaks became reachable, together with new forms of making a living. Once again, islands tell this story best. They are not only graveyards of extinct species, but also wellsprings of evolution's creativity.

On the tiny Galápagos Islands, a single founding colony of finches diversified into fourteen different species, some of which Charles Darwin discovered when he visited on the HMS *Beagle* in 1835.[28] On Hawaii, at least thirty species of nectar-feeding honeycreepers evolved, and on the Canary Islands off the African west coast, twenty-three new plant species appeared in the genus *Echium*—relatives of the blueweed, a modest flowering plant with an eye-catching blue inflorescence.[29] More than 90 percent of one thousand species of flowering plants and more than 98 percent of five thousand species of insects found today on Hawaii have emerged there.

Even more remarkable than these numbers is the explosive speed at which evolution created them. The oldest islands in both the Galápagos and Hawaii have been around for barely five million years, about the same time that separates humans from chimpanzees—a brief moment in evolutionary time that sufficed to create thousands of new island species.[30] But nature's creativity is not just about speed and the number of species. Many new island species also have new lifestyles.[31] The first finches on Galápagos fed on soft insects, but some of today's species have evolved oversized nutcracker-like beaks to crush the hardest seeds to be found. On the Canary Islands, some relatives of the modest blueweed have evolved into eighteen-foot-high wooden giants supported by a drought-resistant root system and crowned by a gaudy cylindrical inflorescence beloved by gardeners.

Some lifestyles have been invented repeatedly on different islands—plants, for example, turned woody more than once—whereas others are one-of-a-kind.[32] Among these unique lifestyles is the feeding habit of the vampire finch, which pecks the tail of blue-footed boobies to feast on their blood. Or take the remarkable skill of the woodpecker finch, yet another peculiar Galápagos inhabitant. Evolution has taught it to use tools like cactus spines and twigs to help it scare insects from their hideouts inside trees. Another example is the split-jaw snake from

Round Island near Mauritius. Its upper jaw is not a rigid bone like ours, but rather it is hinged by a flexible joint, which allows the snake to devour large lizard prey. And there is the marine iguana from the Galápagos, the only lizard alive today that can live and forage in the sea. Among its multiple innovations are glands to extrude excessive sea salt from its body. And the life-style of some Hawaiian moth larvae from the genus *Eupithecia* couldn't be further from that of harmless leaf-munching caterpil-lars.[33] They are the stuff of horror movies. Disguised as leaves or twigs, they assassinate insects landing near them, snatching their unsuspecting victims lightning-fast with specialized, pincer-like legs.[34]

All these are a smattering of nature's myriad creative solu-tions to the same problem: how to survive and make a living when you get stranded on an island.

Through bursts of innovation, islands teach us that creativity can blossom when competition's grip is not too powerful. How-ever, they showcase only what's possible in a short interval of evolutionary time, when genetic drift meets new and empty is-land environments. That's because these innovation bursts on is-lands usually unfold over a mere few million years, a tiny fraction of the four billion years that have passed since life's origin.

In the unimaginably large time span since that origin, some-thing even more profound than the creation of unusual island faunas has taken place. The power of drift has increased slowly but steadily as evolution has created ever more complex and larger organisms. With this increasing power, genetic drift has transformed the very genetic substrate that enables nature's cre-ativity: it has fashioned genomes whose architecture is primed for innovation. Here is how.

A hundred square meters can host a population of ten trillion microbes, whereas even a hundred square *kilometers*—a million times as much space—is barely enough for a handful of large mammals, some forty lions, fifteen tigers, or two polar bears.[35] The larger an organism is, the more space it needs. And this also means that larger organisms generally live in smaller populations. As a rule of thumb, bacteria live in groups of some one-hundred million individuals, small invertebrates like insects or some worms cohabit with ten million others, and the populations of vertebrates and trees typically lie below ten thousand members.[36] Population sizes vary, of course, within these groups—elephants and mice are both vertebrates, but our planet surely hosts more mice than elephants. However, the general trend—larger organisms, smaller numbers—is clear, and so are its consequences: in the smaller populations of larger organisms, the influence of genetic drift is greater, and that of selection is weaker than in larger populations of smaller organisms. (If our billion-strong population of humans seems an exception to the rule that large organisms live in small populations, it is worth remembering that the human population was below a few million for most of our evolutionary history. What is more, during this time the human population was subdivided into many small, isolated tribes. Our population has exploded so recently that the evolution of our bodies and our genomes has not yet caught up.[37])

In the life of large organisms, drift is ten thousand times stronger—and selection that much weaker—than in tiny bacteria with their huge populations. Many bad alleles that would be wiped out quickly in a bacterial population are invisible to natural selection and can persist in large animals or plants. This increasing power of drift in larger and more complex organisms has several surprising consequences. Most important, it has allowed the size of our genomes, the number of its DNA letters, to creep upward, steadily, over many millions of years.

A typical vertebrate genome like ours has three billion letters, almost a thousand times as many as that of a bacterium like *E.coli*. And while we need more genes to build and maintain our complex bodies, we don't have *that* many more genes—less than seven times more than *E.coli*'s four and a half thousand. In other words, our greater number of genes cannot explain why our genomes are so much larger. Most of the difference in genome size comes, in fact, from the DNA *outside* of genes. Such DNA is also called *non-coding* because it does not encode any proteins.

To be sure, the genome of a bacterium like *E.coli* also harbors a bit of non-coding DNA. Much of it consists of short DNA words necessary for gene regulation. These words are recognized by the protein regulators of transcription, which I first mentioned in Chapter 2. A regulatory protein can latch onto such a word and turn on or shut off a gene's transcription—the essential prelude to protein synthesis. In other words, most of *E.coli*'s non-coding DNA has a specific purpose. It helps regulate genes.[38] The amount of this DNA is quite modest, covering only about 12 percent of *E.coli*'s genome.

Our genome couldn't be more different. It is a vast sea of non-coding DNA in which our genes are but small islands that occupy a mere 3 percent of the genome. Two human genes can be separated by thousands or millions of non-coding letters, whereas two average *E.coli* genes are only separated by some 120 non-coding letters.[39] What is more, only a tiny fraction of human non-coding DNA regulates genes. We do not know yet what—if anything—most of the rest is doing, but as we shall see, it is a giant playground for evolution's creativity.[40] Genetic drift is crucial to understanding where it is coming from.

As a genome is passed on from generation to generation, its size can increase in several ways. One of them is DNA duplication, a kind of DNA mutation that happens no less frequently than the single-letter changes—point mutations—we encountered earlier. It occurs when cells aim to repair damaged DNA and commit a particular kind of error, a bit like an editor who proofreads an electronic manuscript and copy–pastes a paragraph of text by error. These errors are not rare, because DNA constantly gets damaged, and cells thus incessantly edit their DNA.[41]

The DNA text copied in a duplication may comprise a few letters, thousands of letters, or large parts of a chromosome with millions of letters. It may also comprise one or more genes. In this case, a gene duplication has occurred.

A duplicated gene is exposed to the same constant drizzle of DNA-changing mutations that falls onto the rest of the genome. If it is lucky, one of these mutations will teach it a new trick— perhaps it can help digest a new kind of food molecule or defend the cell against some toxin. But most of these mutations do what mutations usually do: muck things up. They impair or destroy a gene's ability to make a useful protein. The mutation then has turned the duplicate into a stretch of inert DNA known as a pseudogene, and when a pseudogene is born, a genome's content of non-coding DNA has grown.

To understand the fate of duplicated DNA in evolution, consider that duplication is not free. It costs energy. That's the energy a cell needs to manufacture the building blocks of the duplicated DNA. And if that DNA includes one or more genes, it also requires additional energy to decode the DNA's information and manufacture the encoded protein. In a 2007 study, I used data from complex experimental measurements of this energy to calculate that a gene duplication typically consumes some 0.01 percent of a microbial cell's energy budget.[42] That energy is no longer available for other purposes, such as reproduction.

A hundredth of a percent does not sound like much, and indeed it escapes the notice of our scientific instruments. But natural selection is a more discriminating judge. Recall that microbes live in huge populations, where tiny differences in fitness matter. A microbe hosting such a duplication would be slowly outcompeted by its more efficient brethren. That might take thousands or millions of generations, but eventually and inexorably, its descendants would disappear.

Most animals and plants with the same duplication would not. They live in smaller populations where drift is stronger and the same kind of difference is invisible to selection.[43] As a result, duplicated DNA can accumulate, one by one, millennium after millennium, during the unimaginably long time evolution needs to unfold. The end result? Vast stretches of non-coding DNA. They include some fifteen thousand pseudogenes that litter our genomes, and probably untold thousands more whose features have been washed out beyond recognition in the steady rain of DNA mutations.[44]

Genes become duplicated passively, through no effort of their own, but another kind of DNA—mobile DNA—aggressively promotes its own duplication. A stretch of such DNA is usually a few thousand DNA letters long, and it encodes one or more proteins with a peculiar talent: the ability to copy and paste their coding DNA to some, usually arbitrary location elsewhere in the genome. From there, this DNA can get copied again and again, ad infinitum.

Mobile DNA is a quintessential example of what Richard Dawkins described in *The Selfish Gene*.[45] Not serving any higher purpose, it multiplies within its host's genome without regard for the host's well-being. If nothing were to hold mobile DNA in check, a genome could become overrun by its copies.[46]

Fortunately, something does keep mobile DNA in check, and that something is natural selection.[47]

If you were to paste an arbitrary paragraph of a novel's text into a random new location, chances are that the novel would get worse (unless it was awful to begin with). By the same token, bad stuff also happens when mobile DNA inserts itself haphazardly into a new genomic location. When it gets pasted into another gene, it can disrupt the gene's information string and damage the instructions to make a useful protein. If that gene is required, for example, by a developing embryo, the result may be the embryo's death. Also, when mobile DNA gets pasted near a gene, it can inadvertently turn that gene on. That's because mobile DNA contains regulatory sequences necessary to turn on its own genes for relocation, and this regulatory DNA can activate any gene that happens to be nearby. When this happens to a gene involved in an embryo's development, and the gene gets turned on at the wrong time and place, development can be altered dramatically or subtly—two nerve cells do not connect properly, a blood vessel does not form in the right place, or a bone is not quite as strong as it needs to be. Indeed, subtle changes are more frequent than dramatic ones, and so the damage is often slight, reducing the host's fitness by less than 1 percent.[48] And while natural selection will quickly eliminate any insertions that wreak havoc, selection alone does not decide the fate of insertions with subtle effects. Genetic drift may also have a word to say.

In organisms with huge populations, like *E.coli*, the influence of drift is weak, and selection has no trouble wiping out most damaging mobile DNA insertions. That is why most microbial genomes contain little mobile DNA—usually 1 percent or less of their genomes.[49] In large organisms, however, mobile DNA can steadily accrue because populations are too small, genetic drift is too strong, and selection is too weak to weed out mobile DNA with subtle effects. The end result: more than 50 percent of our genome—and that of other large animals and plants—has mobile origins.[50] Our genomes contain millions of copies of mobile DNA.[51]

And while mobile DNA has been steadily accumulating in our genome, it was exposed to the same rain of mutations as all the rest of our DNA. Because these mutations can destroy its ability to copy and move over time, the vast majority of our mobile DNA—more than 99 percent—is crippled and inert.[52] It can no longer relocate, and its genes have been washed out to become pseudogenes. Even when crippled, though, it still contributes to the ocean of non-coding DNA in our genomes.

The upshot is that larger organisms generally have more complex genomes, courtesy of genetic drift. As organisms get larger, they live in smaller populations—selection becomes weaker and genetic drift stronger—and their genomes acquire more genes, more duplicate genes, more pseudogenes, more active mobile DNA, more defective mobile DNA, and overall more non-coding DNA, so much of it that their genomes' size has steadily increased more than a thousand-fold.

Indiana University biologist Michael Lynch was among the first to show that genetic drift is important to explaining our ever-increasing genome complexity. Lynch made his case by comparing the genomes of hundreds of different organisms, and his data reveals not only that drift helps increase genome size, but also that drift increases the complexity of individual genes.[53]

When the DNA of some genes is transcribed into RNA, parts of these genes, known as *introns*, are eliminated, while other parts, known as *exons*, are joined or spliced together. Only the joined exons are translated into proteins. In other words, genes can come in pieces that are assembled only when the genes' information is decoded.[54] During life's ascent, the number of these pieces per gene rose steadily, just like genome size did. Whereas a typical gene in a microbe comes in one or two pieces, the genes of mice and men have more than seven.[55] What is more, the size of the discarded intronic DNA steadily increased, such that in our genomes, more than 98 percent of a gene's transcribed DNA is discarded, and less than 2 percent is translated into protein.[56]

The result of all this genomic complexity is a giant playground for evolution's creativity. The more pieces a gene has, the more kinds of proteins can be created by mixing and matching these pieces in new ways. Our genes come in so many pieces that our body can make fifty thousand more proteins than a fruit fly can, even though we have fewer than twice as many genes.[57]

In addition, because typical vertebrate genes are separated by thousands or millions of non-coding DNA letters, random mutations stand a far greater chance of creating new DNA words that can be bound by regulator proteins and bring about new gene regulation in genomes like ours than they do in *E. coli*, with its minuscule amounts of non-coding DNA.[58] That's why changes in gene regulation have been crucial to the evolution of large and complex organisms—perhaps more so than changes in genes themselves. Most genetic differences between humans and chimpanzees, for example, are changes in non-coding DNA that can alter gene regulation.[59] Such changes modulate life's recipe in ways that appear subtle but that can have effects as dramatic as creating a new species capable of symbolic language, art, and literature.

Unlike the genome of a microbe, which is like the spartan, barely furnished cell of a monk, the genome of a multicellular organism resembles the workshop of an inventor, filled to the rafters with spare parts, tools, abandoned projects, disassembled machines, and half-finished designs—in short, the kind of junk that is the seed of the next breakthrough invention. And central to filling this workshop with useful parts is genetic drift. As long as our inventor can keep natural selection—an overzealous janitor—in check, he can keep on tinkering.[60]

In sum, genetic drift affects life's evolution in two fundamental ways. In the short run—the few million years needed to form new species—it helps evolution attain new and higher peaks in nature's genetic landscapes. In doing so, it accelerates the creation of new species with unique lifestyles. And in the long run, drift alters the architecture of genomes and increases their potential for future innovation.

As different as these manifestations of genetic drift are, they share a common principle: good things can happen when evolution is free to explore the landscapes of nature's creativity, temporarily liberated from selection's shortsighted and relentless uphill drive.

As it turns out, genetic drift is not the only help evolution has in traversing these landscapes.

Chapter 4

Teleportation in Genetic Landscapes

As difficult as it is to imagine traversing a mountain range in four or more dimensions, I can think of something that is even more difficult: traversing a mountain range in only two. The problem in two dimensions, however, is not our lack of imagination. Darwinian evolution would be much harder in two dimensions because a flat mountain range is more difficult to traverse than one in three or more dimensions—far more difficult.

Imagine that the upper part of Figure 4.1 shows a fitness landscape in a flat world. A population that lives on the left plateau and needs to get to the right plateau would face an already familiar problem: natural selection would prevent the population from traversing the valley separating the plateaus because it forbids even a single downhill step. But what if that figure were really a two-dimensional slice through a three-dimensional landscape, the profile of a crater like that shown in the lower part of Figure 4.1, shaped, perhaps, by a meteor strike? The two plateaus in the flat plane would then become a single three-dimensional plateau, really a circular ridgeline, undulating in altitude, perhaps, but easily circumnavigated by a population propelled only by the modest amount of genetic drift present even in large populations. No need for the strong drift—and tiny population—required to go through the crater's bottom.

Although our three-dimensional brains cannot visualize it, the same idea works in higher dimensions. If the Rocky Mountains, or the Alps, or the Andes were just a three-dimensional profile of a mountain range in four dimensions, some of their peaks—separated by deep three-dimensional valleys—could

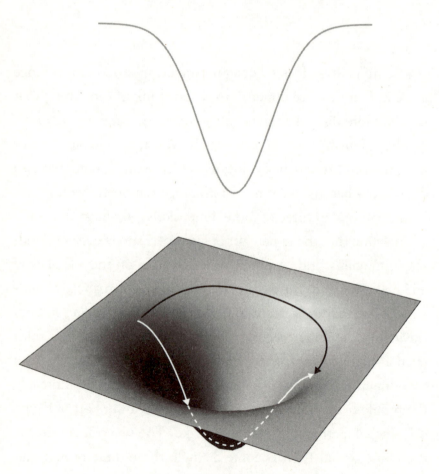

Figure 4.1.

be connected by a ridge in the fourth dimension. And some of those peaks that were still inaccessible in four dimensions could be accessed via an easy stroll in five dimensions, leaving fewer inaccessible peaks, which could become accessible in a sixth dimension, and so on.

Fitness landscapes, of course, do not exist in only three, four, or five dimensions: they exist in hundreds and thousands. Each dimension corresponds to one feature of an organism that can vary, such as a location in the organism's genome—be it a gene or an individual DNA letter—that can vary in its DNA text. For many purposes, thinking of such landscapes in three dimensions works well, but a paltry three dimensions fail us when we try to understand the ability to bypass fitness valleys. Fortunately, even though our brains cannot visualize high-dimensional landscapes, we can still explore and map them.

My research group in Zurich and many others are doing just that, using state-of-the-art laboratory evolution experiments and computing technology. In our experiments, we evolve RNA enzymes and protein enzymes, like the beta-lactamase that neutralizes the antibiotic penicillin. To do this, we manufacture huge populations of identical, high-functioning molecules. We then sprinkle a population's individuals with mutations in their molecular text, select for mutants that still perform well, and repeat. And while we do this, we sequence the molecular texts of the evolving individuals—thousands of them—and ask our computers to track their location on the landscape.

What such populations teach us is just as important for evolution's creative powers as it is surprising and bizarre.

In these experiments, any one population starts out at a peak of its adaptive landscape. If this peak were isolated on the landscape's high-dimensional plains, like Africa's Mount Kili-manjaro, the population's member molecules would remain huddled around the peak. All mutants would be inferior to the starting molecule, and natural selection would relentlessly purge them from the population. But that's not what we see. The population does not stay put. It spreads through the landscape. Inexorably, steadily, and with every cycle of mutation and selection, its members take a few further mutational steps. What is more, individual members move away from the starting peak in many different directions. While they move, they get neither much better nor much worse at what they do, staying at about the same elevation on the landscape. The starting peak, it seems, is connected to other, nearby peaks. Dozens of "ridges" emanate from it and lead to these other peaks, from which dozens more ridges connect to even farther peaks, and so on.

Experiments like these teach us something profound about the architecture of multidimensional adaptive landscapes: a peak is usually not a single location, like the top of Mount Kilimanjaro, but more like a network of high-altitude paths that form a sprawling spiderweb extending far through the landscape. The paths connecting different peaks need not be completely flat, but they also do not snake up and down by much. We know this because our experiments use huge populations of many billions of molecules, and in such populations genetic drift is too feeble to allow any one individual to descend far below a peak.

Laboratory evolution experiments can only explore a tiny speck of an adaptive landscape, but in its four-billion-year experiment, nature has explored much greater swaths. It has also discovered how far molecules can travel on spiderwebs like these while preserving their unique skills. Take hemoglobin, the oxygen-binding transport protein in our blood. Thousands of other species—mice, reptiles, fish, insects, and even plants—harbor oxygen-binding proteins like it, each of them the endpoint of a long evolutionary journey that started from some ancestral oxygen-binding protein more than a billion years ago. And in this journey, oxygen-binding proteins have not remained unchanged. Their amino acid texts have changed slowly and inexorably, letter by letter by letter. Starting from their ancient common ancestor, these proteins have spread out on a vast spiderweb of useful oxygen transporters. By now, they have traveled so far that they share fewer than fifteen of their hundred-odd amino acid letters. They have become very different texts, different solutions to the same problem of binding oxygen.[1]

And hemoglobin shares this pattern of evolution with myriad other evolving molecules, from the catalytic holdovers of a sunken RNA world, to innumerable proteins that catalyze biochemical reactions, communicate between cells, support our bodies, and help us move. Whatever each of them does, it does it very well. And the fitness peak each occupies is not merely a single peak, but rather a sprawling network of high-dimensional paths that extends far and wide through an adaptive landscape. Don't try too hard to visualize this spiderweb, because its very existence would be impossible in three dimensions. It requires hundreds of dimensions and the many directions in which they can be explored.

What is more, all this applies not only to individual molecules, but also to complex assemblages of these molecules—the biochemical machineries that build and maintain our bodies. Two of these machineries are especially important. One is run by the regulator proteins that control the transcription of many genes. These regulators do not act alone, but instead in complex regulatory circuits whose members regulate both each other and the transcription of hundreds of other genes. The second machinery is metabolism, a complex network of thousands of chemical reactions, each catalyzed by a dedicated enzyme encoded in some gene. Our metabolism and those of all other organisms procure energy and building materials—nutrients—to manufacture the numerous molecules life needs to persist.

These regulatory and metabolic machineries are encoded in the genome, and when the genome varies through DNA mutations, so do they. This is why a population's individuals differ in their metabolism, some harvesting energy more efficiently than others, some storing more fat, some tolerating certain foods better, and so on. And this is why some regulation circuits build better bodies than others, constructing larger wings, stronger hearts, or faster neurons. In other words, regulation and metabolism can evolve and improve by DNA mutation and natural selection. But just as important is this: even the circuits and metabolisms achieving peak performance are not all the same. The peak they occupy is not a single Kilimanjaro-like hump in the landscape. Rather, it is a sprawling network of ridges. Along this network, diverse forms of regulation and metabolism can co-exist. They are each able to build and maintain an optimally functioning body but do so in different ways.

Entire books could be written about this unexpected high-dimensional world. In fact, one has been written. In *Arrival of the Fittest*, I tell the fascinating story of where these spiderwebs come from, why they are near-universal, and why they are crucial for evolution.[2] But what matters here is that they help nature in the act of creation and that we have the experiments to prove it.

In one of these experiments, Eric Hayden, then a postdoctoral researcher in my laboratory, started with a ribozyme (already mentioned in Chapter 2) that can link an RNA molecule with a specific letter sequence to itself. From this starting point, Eric sought to evolve a more flexible ribozyme that could link a different RNA molecule to itself. Actually, he performed two experiments, each starting with a different population of molecules. The first population was concentrated on one peak of the starting ribozyme's adaptive landscape—actually, in one location of a high-altitude network of ridges—whereas the second was sprawled out along this network. He asked which of the populations would be the better innovator—which would discover the new molecule faster.

After a mere eight cycles of mutation and selection we had the answer. The population spread far and wide discovered the more flexible ribozyme six times faster. And the reason is not hard to understand: this new ribozyme occupied a higher peak that existed some distance away from the starting ribozyme, and in the population spread out across the ridges below this new peak, some individual ribozymes just happened to be close to the new peak. They had a head start and could get to the top faster.[3]

Much the same story is told by the antibiotic resistance protein beta-lactamase. An experiment in the laboratory of Israeli biochemist Dan Tawfik showed that evolving beta-lactamase genes that are spread out along ridges near their adaptive peak have better odds of developing the novel ability to destroy both cefotaxime and penicillin. Different molecule, same reason: the sprawling network can help them get close to a new peak without their dipping into a deep valley.[4]

That very fact can help explain why resistance against ever-new antibiotics evolves so rapidly. Not only do bacteria divide fast and live in huge populations, but their antibiotic resistance genes are also highly diverse and scattered all over the adaptive landscape, thanks in part to the sprawling spiderwebs of ridges permeating this landscape.[5] This diversity increases the odds that one of these genes will find itself near a peak of resistance against a new antibiotic.

In sum, the peculiar architecture of adaptive landscapes, where each peak really is a network of multi-dimensional ridges, helps evolution solve difficult problems. It helps create diverse organisms and molecules, some of which may be close to even higher peaks—better solutions to old problems or inventive solutions to new problems.

"Beam me up, Scottie" is an indelible line etched into the collective memory of those who are familiar with the original *Star Trek* television series.[6] James Kirk, the captain of the starship *Enterprise*, spoke it whenever he needed to get himself out of a tight spot, usually among hostile creatures on an alien planet, whereupon the starship's engineer, Montgomery Scott, would magically teleport Kirk back onto the mothership. Sadly, like the *Enterprise*, this kind of teleportation remains a matter of science fiction.

At least in our daily lives. Because it turns out that nature uses something like teleportation to reach faraway places—not quite a starship, but distant locations on the landscapes where nature's creativity unfolds. Everybody knows about it, most people like it, but many fewer truly appreciate why it is important: sex.

Each of the twenty-three chromosomes that host our genes comes in a pair—that's why we have two copies of each gene. During a special kind of cell division that creates sperm and egg cells, a chromosome pair's two members line up to exchange some of their DNA text. Imagine each pair as two equally long shoelaces in different colors, say, black and white, lined up in parallel on a flat surface, end to end, one on the left, and the other on the right. While keeping these shoelaces aligned, cut them in an arbitrary place—actually, cut them in a few places—and then swap the resulting fragments between the left and right. When you are done, glue the pieces back together, and you are left with a pair of shoelaces whose members change color—black to white or white to black—at least once along their length. Wherever the left shoelace changes from black to white, the right one changes from white to black.

When our bodies manufacture sperm and egg cells, this aligning, random cutting, swapping, and regluing—the scientific term is *recombination*—happens to each chromosome pair, which becomes a mosaic of DNA strings, just like those rearranged shoelaces become a mosaic of black and white strings. And one member of each pair gets stuffed into each sperm or egg cell.

When two parents conceive a child, the sperm spills the father's rearranged DNA into the female egg, which contains the mother's rearranged DNA. The end result is a fertilized cell that has two copies of each of the twenty-three chromosomes that were recombined inside the mother and father.

In reality, the two DNA strings differ not in their color but in the sequence of their DNA letters, and only ever so slightly, at about one in every one thousand letters for a typical human being.[7] That is, if you were to walk along any one of your chromosome pairs, you would find that once every one thousand letters, one of the pairs has one letter, say, A, whereas the other has another letter, C, G, or T. In all the remaining letters, 99.9 percent of them, the two chromosomes are identical.

In other words, the two members of each of your twenty-three chromosome pairs barely differ. But because there is so much DNA in them—some three billion letters, packed into all those chromosomes—these differences add up, such that, overall, three million DNA letters differ between the two copies of all of your chromosomes.[8]

Knowing this, we can also find out how many letters differ between the newly assembled genome of a child and the genome of either parent. One member of each chromosome pair comes from the father, so it does not differ from that father's genome.[9] The other one comes from the mother, which differs from the father by about three million letters. Overall, the child's genome thus differs from the father's genome by some 1.5 million letters— the average of zero and three million. By the same calculation, the child's genome differs from that of its mother by the same amount, some 1.5 million letters, or 0.05 percent of its genome.

This percentage may not sound like much, but adaptive landscapes can help us grasp its true magnitude. If a single step on an adaptive landscape—a single letter change in a genome— covered as much distance as an average human covers in a single step, then this kind of genome swapping would teleport a child about seven hundred miles, traversed in a single leap. And genomes take such leaps every time two parents have a child.[10] If you were to travel this far from the rolling plains around Wichita,

Kansas, you could find yourself in the middle of the Rockies in Colorado or Utah, with plenty of new peaks to explore.

The DNA of human parents—or any two organisms of the same species—usually differs much less than does the DNA of different species. But the more two parents differ in their DNA, the further recombination can leap, and the greater its creative powers can become. Recombination leaps especially far when parents from different species mate and produce offspring—the kind that is called a species hybrid. To be sure, some hybrids are dead ends of evolution—their parents are very different or genetically incompatible, so a fetus cannot develop, or it develops but is sterile. Examples include the mule, a horse–donkey hybrid; the zorse, a zebra–horse hybrid; and the liger, a lion–tiger hybrid. But hybridization can also be very successful. It can even launch entirely new species—instantly.[11]

Successful hybridization is especially common in plants, where it creates up to 10 percent of new plant species.[12] It also often enables the new species to boldly go where no parents have gone before. Case in point: two hybrid species of US sunflowers from the genus *Helianthus*. Their parents dwell on the Great Plains, but one of the hybrid newcomers can survive in the deserts of Nevada—it is aptly named *Helianthus deserticola*. The other thrives in Texan salt marshes. Both new habitats would be deadly for either parent.[13]

Animals, too, can hybridize successfully. In 1981, for example, Princeton researchers Peter and Rosemary Grant discovered a new hybrid Galápagos finch when they encountered an especially unusual male specimen on the island of Daphne Major. Not only was this bird 50 percent heavier than other finches, it also sang a new song, had an unusually large head, and had a beak that allowed it to crack seeds that were inaccessible to other finches. By observing "Big Bird's" descendants through seven generations during the next twenty-eight years—talk about tenacity—the Grants found that its new features were indeed helpful. When a drought wiped out 90 percent of finches between 2003 and 2005, its descendants were among the survivors. DNA analysis showed that Big Bird was a hybrid of two other Galápagos finch species. More than that, it proved that multiple other species of Darwin finches are hybrids.[14]

Bacteria cannot hybridize like plants and animals do. However, teleportation across adaptive landscapes is so important that nature empowered even bacteria to do it, though it equipped them with a mechanism very different from ours. Bacterial genomes can include genes that enable one bacterium to donate DNA to another, recipient bacterium. These genes enable the donor to build a long, hollow protein tube—the technical term is a sex pilus—that latches onto a nearby recipient cell, reels it in, and helps transfer DNA to the recipient. The process is also called horizontal gene transfer, and the transferred DNA can help the recipient survive in new environments. Horizontal gene transfer may seem vaguely similar to our version of sex, but it differs in important respects. Bacteria do not have sex every generation, nor do they reproduce with sex like we do. They merely transfer a copy of DNA—up to hundreds of genes—to another cell. Sometimes the process even transfers the very genes needed to build the sex pilus. The result is something like a sex change that converts a "female" into a "male."[15]

But the most important difference from our sex is this: bacterial sex is available to organisms a hundred-fold more diverse than two humans or even two sunflowers.[16] If we had a bacterium's recombination powers, we would not just routinely blend genes from other humans into our genome, but also those from chimpanzees, mice, birds, or even reptiles and fish. Bacteria can even exchange DNA with animals and plants.[17] Just imagine the consequences—on our lifestyle, on world hunger, and on the global economy—if we could acquire a plant's ability to harvest energy from light and to build our bodies from the air's carbon dioxide.[18]

In sum, bacteria can catapult themselves not just hundreds of miles, but thousands of miles, through a vast genetic landscape, all courtesy of gene transfer.

It is not surprising then that bacterial teleportation has led to innovations almost as radical as humans capable of photosynthesis, because bacteria all around us exploit teleportation to experiment with ever-new combinations of genes. Among their creative discoveries are gene combinations that help them survive—and even thrive—on nasty man-made molecules, such as pesticides like DDT and pentachlorophenol, or the highly toxic industrial waste product dioxin.[19]

Molecules like these were invented by chemists only within the last century, which gives you an idea of how quickly bacteria created the innovations that turned such molecules into food: a mere instant of evolutionary time. And once gene transfer has helped create any one such innovation, it helps spread the innovation from one bacterial species to another and beyond. That's why the skill of surviving an antibiotic can spread rapidly among different species, so rapidly that human innovators have trouble creating new antibiotics fast enough to catch up.[20]

Fortunately, nature's lessons about distant jumps through an adaptive landscape are not lost on human innovators. They are imagining new mechanisms for genetic teleportation that are more powerful than even bacterial sex. These strategies create new DNA in test tubes through an orgy of molecular recombination that even nature would be hard-pressed to match.[21]

Among these innovators was the late Pim Stemmer, a Dutch biochemist and serial entrepreneur with dozens of patents to his name. Stemmer's fame in biotech circles comes from his 1994 invention of DNA shuffling, a biochemical technique that uses an enzyme called DNA polymerase to copy huge numbers of DNA molecules that encode one or more genes.

DNA polymerase is by no means an exotic enzyme, a laboratory creation of biotechnologists. It occurs in every living cell and is essential to making a copy of a cell's DNA whenever the cell divides. However, biotechnologists use an engineered form of this enzyme and exploit a property that is crucial for DNA shuffling.[22] Starting from one end of a DNA string, DNA polymerase glides along the string and copies it letter by letter. While it does, it can hop from that string to another nearby DNA string—biochemists say it switches templates—and continue copying the other string. It's as if you had two similar English texts side by side, and sometime after you started copying one of them, you moved over to the other text and continued copying that one. The polymerase's jump makes no difference to the copy's letter sequence if the two DNA strings are identical, but if they are different, the resulting copy is a chimaera, starting with the letter sequence of the first DNA text and ending with that of the second.

With DNA shuffling, biochemists can shuffle mixtures of many molecules, each with a different letter sequence. The DNA they shuffle can also be highly diverse, more so than the typical genes of two amorous bacteria—it could come from organisms as different as marmots and marigolds. What is more, while replicating any one of these molecules, polymerase can switch templates multiple times, such that the final copy contains text snippets from multiple DNA strings.

Think of DNA shuffling as group sex for molecules.

The long jumps through the adaptive landscape enabled by DNA shuffling proved their mettle when researchers in Stemmer's laboratory tested the power of DNA shuffling to improve on nature's handiwork in creating efficient enzymes.[23] They focused on one of those enzymes that disarms antibiotics like cefotaxime, starting with a pool of genes from four different bacteria that encoded different variants of the enzyme. Very different variants, I should add: up to 40 percent of the letters in their DNA text differed among them. To begin, the researchers asked how much they could improve the starting enzymes by walking through the landscape one step at a time, changing individual DNA letters one by one. The answer: eight-fold. In other words, the resulting enzyme could cleave eight times more antibiotic molecules in the same amount of time. Not too bad, you might say. But it is nothing compared to what happened when they forced the molecules into a foursome of DNA shuffling. That foursome created an enzyme five-hundred-fold better than its parents. Other experiments with DNA shuffling created enzymes that remove dirt from clothes faster, cleave new kinds of molecules, or detoxify arsenic-laced mining waste.[24]

DNA shuffling, promiscuous bacteria, and species hybrids teach us that teleportation in genetic landscapes is crucial to nature's creative powers. We would therefore expect to find it everywhere on the vast tree of life, and that's indeed the case. Almost. Some of the tree's million-plus species apparently do not recombine their DNA. These include some salamanders whose females propagate through unfertilized eggs, and flowering plants whose seeds develop without a pollen grain.[25] But what's telling about these species is that almost all of them form tiny twigs on life's tree. No major branch of animals or plants reproduces asexually. That observation, as innocuous as it seems, makes a profound statement about the importance of sex. Asexual species did not just lose sex, they lost it recently in their evolutionary history, otherwise larger branches of life's tree would be sex-free. Species that lost sex many millions of years ago are no longer around. They suffered evolution's ultimate death penalty: extinction.

The message could not be clearer: lose sex and you are not long for this world.

But here is a mystery, an apparent exception to this rule—a tiny fraction of species known as ancient asexuals. They seemingly made do without sex for millions of years. Among them are some three hundred species of tiny freshwater animals called the bdelloid rotifers, which originated more than thirty million years ago.[26] While no amount of searching turned up any evidence for hanky panky in these critters, a recent analysis of their genome's DNA revealed something even more remarkable than their apparent asexuality: more than three thousand genes in their genome are not their own.[27] These genes do not even come from other multicellular animals. They have been transferred into their genome from who-knows-where.

We have no idea how the bdelloid rotifers do it, but they clearly utilize the same kind of horizontal gene transfer perfected by bacteria to leap through an adaptive landscape. In other words, these asexuals are not so asexual after all, even though their teleportation machinery still awaits discovery. And perhaps other ancient asexuals are like them? Perhaps they also are secretly sexual, harboring genomic signatures of unusual sexuality? It's good to know that in the twenty-first century there are still biological mysteries to be solved and important discoveries to be made.

Recombination's near-universality testifies that genetic teleportation has been essential to life's ascent. But it also raises a vexing question. Why does Captain Kirk always land on the mothership and never in outer space? Why do recombination's blind leaps rarely end up in some deep valley of the landscape, producing a mashed-up genome and a broken organism? Or perhaps they do, and the teleportation machine kills most of those who enter it? Regrettably, that's not easy to find out, because the organisms whose genomes leapt off a cliff were never even born, so we cannot examine them.[28] But we can do something else. We can use computers to simulate the long-distance jumps of recombining molecules and genomes.

Different researchers around the world use computers to do just that, and they are coming up with similar answers. Among them is Allan Drummond from the University of Chicago, who has asked where genes land in an adaptive landscape after recombination—closer to a peak or nearer to a valley? More precisely, he has asked whether recombination leaves the protein encoded by these genes unharmed. Likewise, some researchers in my laboratory study recombination in the DNA that encodes the chemical reactions of metabolism. They predict whether a metabolism can still support life after a long-distance jump through its adaptive landscape. Yet other researchers study recombination among the regulators and circuits that help build new bodies. They ask whether recombination leaves the sophisticated recipes for creating whole organisms intact.

All of these scientists compare the harm from the long-distance jumps of recombination with the harm that comes from traveling the same distance, but like a pedestrian, through many steps of random single-letter changes.[29] And they all come up with similar answers: recombination is much more likely to preserve life—up to thousands of times more likely—than random mutation is. To be sure, recombination does have the potential to destroy—just think of those sterile hybrids. However, this destructive potential is much smaller than that of random mutation. It is no match for recombination's enormous creative potential.

The reason? When nature recombines genomes, and when biotechnologists recombine molecules, they do not make completely haphazard changes to DNA. Instead, they take organisms or molecules that already work well—we know, because they have survived to this day—and mix up their parts. It's as if you exchanged the pages of two texts that tell a similar story but in different words. Such recombination will not always improve the text, but it will usually not destroy its meaning completely, and could even create unexpected twists or new plotlines. Not so if you just "mutated" the text through millions of typographical errors. You'd be almost certain to garble its meaning.

Another way to understand recombination's potential is offered by the high-dimensional nature of adaptive landscapes and by the spiderwebs of high-altitude ridges. The very existence of these ridges means that long-distance jumps *can* land in a region of high elevation and preserve adaptation. What is required is that they actually *do* land on a ridge. And recombination's reshuffling helps them achieve that soft landing because it recombines parts of molecules that already work well together.

Together with genetic drift, DNA recombination and the sprawl of adaptive ridges counterbalance natural selection's shortsightedness. They temper selection's compulsive ascent of the nearest hill in an adaptive landscape. Whereas drift takes modest steps—downhill as much as uphill—recombination causes giant leaps through such a landscape. And adaptive ridges do not only permit a soft landing after recombination, they also permit diversity in a population, which opens new vistas and enables some individuals to ascend new, even higher adaptive peaks.

Nature has come up with multiple ways to temper natural selection, which tells us how important such tempering must be. As it turns out, the need to temper selection has analogs in the human realm. We have seen hints of it in the mental journey that the physicist Hermann von Helmholtz described using when he solved difficult problems, and we will encounter it again later in multiple forms, such as the meandering lives of eminent creators that allow the cross-fertilization behind scientific revolutions.

Carnivorous caterpillars, desert sunflowers, and toxin-gobbling bacteria are only a few of the myriad organisms that recombination and drift helped natural selection create. Together with the spiderweb ridges of adaptive landscapes, these mechanisms of evolution are essential for nature's creative powers. As we shall see next, these powers are so formidable and far-reaching, they even extend to the inanimate world, where shortsighted hill-climbers don't get very far either.

Chapter 5

Of Diamonds and Snowflakes

Geodesic domes may be the biggest triumph of architecture since the Gothic spire. Buildings like the Montréal Biosphère or Spaceship Earth at Florida's Walt Disney World are built from a latticework of struts that form a hollow cage prized for its lightness and stability. The name comes from the large circles—geodesics—that revolve around a sphere's center and that are traced by the struts. Geodesic domes were invented by the German engineer Walther Bauersfeld after World War I, but they became popular only when the American architect and inventor Buckminster Fuller touted them as a solution to the world's housing problems.[1]

These gravity-defying structures could be Exhibit A for human creativity and its unique powers were it not for an annoying little fact: light years away, in the infernal cauldron of ancient stars and nebulae, nature has been churning out miniature versions of them for eons.

We have to thank chemist Harold Kroto and a team of collaborators for this discovery. In 1985, they were puzzled by data

from spectrometers that hinted at the existence of complex carbon molecules with more than a dozen atoms near distant stars
and in interstellar space.[2] Wondering how such molecules could
form in the hostile environment of outer space, they tried to create them in the lab at the high temperatures found near stars.
In a now-famous experiment performed at Rice University, they
shot a focused laser beam at a piece of graphite, which created
infernal temperatures above ten thousand degrees and instantly
vaporized the graphite into atomic carbon.[3]

As these atoms cooled in a jet of helium, they formed molecules even more complex—and beautiful—than those the
scientists had been after. In each molecule, sixty carbon atoms

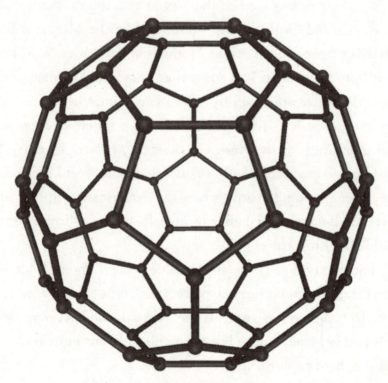

Figure 5.1.

bonded to form a highly regular spherical cage with thirty-two faces—twenty hexagons and twelve pentagons (Figure 5.1). It's the shape of a regulation soccer ball, or, if you like mathematical jargon, of a truncated icosahedron.

In honor of Buckminster Fuller's geodesic domes, Kroto called these molecules buckminsterfullerenes, but they soon came to be known—easier on the tongue and more affectionately—as bucky-balls. Their discovery would win Kroto and two colleagues the 1996 Nobel Prize in chemistry.[4]

Bucky-balls far surpassed the complexity of the interstellar carbon molecules that had prompted Kroto's experiment. But, as it turns out, they do indeed exist in outer space, even though it took decades to discover them there. In 2010, another team of scientists found that bucky-balls assemble by the trillions in carbon-rich shells of old stars and interstellar nebulae.[5] So great are their numbers that they can blot out the light of nearby stars.[6]

Nature created the structure of today's universe long before we came along, and even long before life came along, when the remnants of the Big Bang assembled into atoms, and those atoms assembled into swirling galaxies, whose gas clouds assembled into trillions of suns and even more planets—all by themselves. But nature's creative power is nowhere more evident than in beautiful molecules like bucky-balls. And understanding how such molecules self-assemble holds important lessons about all creativity.

Two carbon atoms bonded together in a buckminsterfullerene—or in any other molecule—are a bit like two tiny balls on a spring. When you pull them apart, you need to expend energy. This energy gets transferred to the spring and stored there in a form physicists call potential energy. Think of it as the atoms' ability—their "potential"—to move closer once you release them. The stronger you pull, the more potential energy they accumulate. And the same thing happens when you push the atoms together: they store potential energy and will release it as soon as you stop pushing and allow the atoms to move farther apart.

Once the two atoms are left to their own devices, they eventually come to rest at some intermediate distance, where their potential energy—or, more precisely, that of the molecule they form—is smallest. That's the lowest point on the parabola of Figure 5.2, which describes the potential energy of a two-atom molecule. Push them together, and they move upward—their

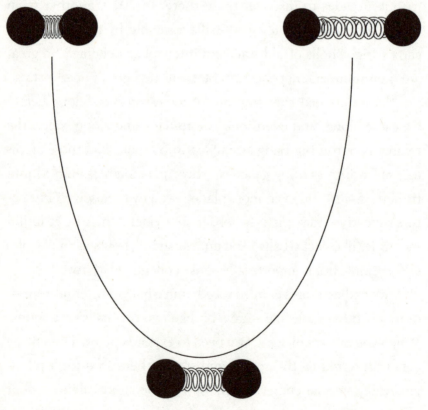

Figure 5.2.

potential energy increases—along the left wall of the parabola. Pull them apart, and they also move upward—their potential energy increases again—but now along the right wall. The harder you push or pull them, the farther they move uphill, and the more energy the molecule stores.

Viewed through a different lens, that parabola is also a simple two-dimensional landscape, so simple that it has only one valley. Chemists call it the potential energy landscape of a two-atom molecule. If you were to drop a marble on the hillsides enclosing the valley, the marble would slide down, roll around near the bottom for a while, and eventually come to rest. The marble's location in the landscape is analogous to the distance between the two atoms. As the marble slides, the atoms move—farther apart or closer together—until they have reached their resting point of lowest potential energy.

Physical laws act on linked atoms like gravity acts on a marble. This principle applies to not just two carbon atoms joined by a *covalent* bond—the kind of strong chemical bond that holds bucky-balls together; it holds for any two atoms and for any kind of physical attraction between them. That includes the attraction between positively and negatively charged ions, like those of sodium and chlorine in table salt. It also includes various kinds of weaker forces that can bond atoms to each other, such as the van der Waals force that supports the three-dimensional shapes of proteins.[7] All these forces are variations on the same theme, like balls of different size linked by springs of different stiffness.

What is more, the same principle applies to more than two atoms. And that's where things get interesting.

To describe your place in the two-dimensional landscape of Figure 5.2, you would need to know only one quantity—the

distance between two atoms. That information would immediately tell you the elevation you are at. It's no longer so simple
when three instead of two atoms are linked. Three balls can be
connected by three springs, which would form a triangle. Each
of these springs can be pushed or pulled, and it can store or release potential energy. In other words, you would need three
numbers—the length of each spring—to describe how far apart
the three atoms are. And you would need a fourth number to describe the potential energy of this atomic configuration. In landscape terms, the first three numbers specify a location, namely that
of the three-atom molecule on its potential energy landscape. The
fourth number specifies the elevation at this location—the molecule's potential energy. In other words, describing a three-atom
molecule already requires a landscape in four dimensions—one
more than in our familiar three-dimensional space.

As the number of atoms increases, the number of springs
increases too, just faster. Four atoms connect via six springs,
five via ten springs, six via fifteen, and so on—the number of
springs increases like the number of possible pairings between
an increasing number of tennis players. And so do the energy
landscape's dimensions. A further complication is that atoms are
usually not confined to a two-dimensional plane like that of Figure 5.2. In the three-dimensional world of bucky-balls, you would
need three numbers to describe the location of each atom. For a
bucky-ball's 60 atoms, 180 coordinates are necessary to describe
the location of all atoms. Combine these 180 coordinates with
the bucky-ball's potential energy—one additional number—and
you have a landscape in 181 dimensions.

Whether a landscape has three dimensions or three hundred,
its topography could be as dreary as the endless plains of Texas or

as simple as the single crater of Figure 5.2. Indeed, up to five atoms, a molecule's potential energy landscape can be that simple: a single valley in which the atoms form a shape called a triangular bipyramid, the only stable molecule they can form.[8] But for six atoms, you get two valleys—two possible stable molecules. For seven atoms, you get six stable molecules; for eight atoms, sixteen stable molecules; for nine atoms, seventy-seven stable molecules; and for ten atoms, 393 stable molecules—each one corresponding to an atomic arrangement where the atoms can come to rest, having expended their potential energy.

The number of stable molecules rises so explosively that beyond ten atoms it quickly becomes impossible to count them all. A mere few dozen atoms can have an energy landscape with billions or trillions of valleys. A few of these valleys are deep, corresponding to the most stable molecules. Their atoms often show highly regular arrangements—cubes, tetrahedra, octahedra (two pyramids glued together at their base), or the truncated icosahedra of bucky-balls. But most valleys are shallow and harbor the least stable molecules, often an irregular jumble of atoms where a small nudge can push all atoms into entirely new configurations.[9]

Potential-energy landscapes of molecules can be rugged, just like the adaptive landscapes where biological evolution unfolds. And just like adaptive landscapes, in which locations correspond to different DNA sequences or genotypes, energy landscapes also exist in an abstract realm with more dimensions than our three-dimensional minds can visualize.

These similarities are momentous. Just like fitness landscapes can teach us how biological evolution creates new kinds of life, so too can energy landscapes teach us how the inorganic world creates the new and beautiful. They can teach us how a jumble of atoms self-assembles into molecules like bucky-balls that are not only complex and easy on the eye, but also so stable that their radiation signature can reach us from other galaxies.

But you may have noticed a difference between the two kinds of landscapes. The high peaks of evolution's landscapes—occupied by well-adapted organisms—are the best places to be, whereas the peaks in an energy landscape are the worst places. They correspond to the most unstable molecules, whose atoms immediately shift their position to release their potential energy, rearranging themselves until they occupy a valley-bound stable molecule.

That difference is less profound than it seems. Consider a raised relief map, the kind of landscape model you find in visitors centers of some national parks like Acadia or Grand Canyon National Park. It is a three-dimensional scale model of the park's landscape that highlights its peaks and valleys. Underneath the surface, such maps are often hollow, and by flipping a hollow relief map by 180 degrees, you turn its peaks into valleys and its valleys into peaks. This simple change in perspective is all you need to convert an adaptive landscape to the energy landscape of a molecule. Where evolution seeks out the highest peaks in an adaptive landscape, atoms and molecules seek out the deepest valleys of their energy landscapes—those corresponding to the most stable molecules.

These landscapes do not just harbor molecules with beautiful architectures; they also contain glittering objects that delight chemical engineers and nanotechnologists. Take the catalytic convertor cleansing a car's exhaust. It contains very expensive metals like platinum and gold whose atomic surface texture can accelerate chemical reactions. These reactions break down toxic molecules such as carbon monoxide, which is how a catalytic converter helps detoxify exhaust gas.

Because a catalyst's surface is so important, equipping a catalytic converter with a solid chunk of gold would be a very bad idea. Most of the gold would be buried inside that chunk. Even in a particle of a mere hundred thousand gold atoms only 10 percent of the atoms would be on the surface. Much better to scatter that chunk into myriad tiny gold particles—gold clusters, as chemists like to call them—so that most of the atoms are on the surface. How big a difference that can make was shown in 2012 by Spanish chemists who increased a gold catalyst's efficiency one-hundred-thousand-fold by equipping it with clusters of no more than ten gold atoms.[10]

Atomic clusters of less precious metals like iron, nickel, or cobalt, together with nonmetals like sulfur or carbon, are important far beyond catalytic converters. They catalyze countless chemical reactions, some of them crucial to the chemical industry, like those creating synthetic lubricants from coal, or fuels from biological waste. Other atomic clusters keep our body going by helping to extract energy from nutrients. Such clusters can self-assemble into various shapes. Their atoms can be spread out like a sheet of dough, bunched up into a ball, or arranged in a crystalline lattice. This shape can make the difference between an efficient catalyst and a sluggish one. To find out whether a catalytic cluster can self-assemble into a shape that is perfect for catalysis, chemists study the deepest valleys in its energy landscape and scrutinize this landscape for obstacles that can prevent this shape from emerging.[11] And when chemists do that, they find a formidable obstacle that is already familiar—the same one that evolution faces when searching the highest peak in an adaptive landscape.

Just as natural selection can only push uphill, gravity can only pull downhill. When atoms of gold, carbon, or iron associate

haphazardly, the result is a jumble, an atom cluster with no organization, corresponding to a marble that is dropped on a random place in a multidimensional energy landscape. Perhaps that
marble lands at a valley or at a peak, but more likely than not it
will come down somewhere on the slope between a peak and a
valley. From there, it will slide downslope to the nearest valley
bottom, where the cluster will find a stable pattern of atoms that
requires the least rearranging. There are many more shallow valleys than deep ones, so this resting place will almost certainly be
shallow—a not-very-stable cluster with its atoms arranged in a
messy jumble. And there the cluster will be stuck forever.

But nature does create bucky-balls, so something must be
missing from this picture. And that something is easy to understand: good vibrations.

The technical term is *heat*, the incessant tremors of atoms
and molecules all around us. The hotter it is, the more strongly
the atoms and molecules will vibrate. When it gets too hot, these
vibrations eventually get so violent that the chemical bonds tying
a molecule or cluster together rupture, and its atoms scatter every
which way. Conversely, at ever colder temperatures, these vibrations get weaker and weaker until they cease completely at absolute zero, a temperature of −273.16 degrees Celsius. In between,
a molecule's bonds—those figurative springs—hold, but they are
incessantly pulled and pushed as atoms are shaken and jostled.
(It's the same kind of jostling that is responsible for the vibrations
of proteins, which allow enzymes to perform useful tasks.)

A metaphorical marble exploring an energy landscape would
jitter unceasingly from these vibrations, as if the landscape itself
were constantly trembling, like an adaptive landscape trembling
from genetic drift. The hotter it gets, the stronger the tremors

become. If that marble started in a shallow valley, even small jitters might help it climb the saddle connecting to a nearby valley, which may be deeper. If the marble started in a deep valley, that saddle would be more like a mountain pass, and the marble could not traverse it unless the tremors were very strong. At the highest temperature where the atoms don't yet fly apart, the tremors would be so strong that the marble would jump around erratically and visit all parts of the landscape, although it would spend most of its time in the deepest valleys—those that are hardest to jump out of.

These jumps help the marble explore the landscape, but they also create a new problem: the marble will never settle in a valley for good—the molecule's atoms will incessantly shift and reconfigure. Fortunately for nature's creative powers, this problem can be avoided by cooling the atoms. Cooling weakens the marble's jitters, so the marble is less likely to leave any deep valley that it is visiting. It will continue to explore that valley, which can itself contain many clefts, cracks, and crevices, and as the temperature drops further, the marble will burrow deeper and deeper, exploring the valleys within a valley, and the even shallower valleys within. If the atoms are cooling slowly enough, the marble will eventually come to rest at the very bottom of the deepest valley, which corresponds to the most stable molecule.

That's at least the world according to theory, the theory of statistical physics—a branch of physics dealing with large numbers of particles like atoms. But this theory works. Ask an amateur chemist laboring to grow large crystals from everyday materials like sugar, salt, or borax in their kitchen, and they will tell you that slow is the way to go—the slower you cool, the larger and more regular a crystal will grow.[12] To be sure, many crystals are not quite like bucky-balls. Their atoms are not bound by strong covalent bonds, but rather by weaker bonds that rupture at more modest temperatures or that dissolve in water, like salt crystals do.[13] What is more, the building blocks of crystals need not be atoms like carbon. They may themselves be molecules like table sugar. But no matter whether it's a sugar crystal, a bucky-ball, or a gold cluster, the principle is the same: particles like atoms and molecules can self-assemble into a stable architecture when each molecular part is free to vibrate and jitter—just enough and not too much—and free to probe myriad configurations of an enormous puzzle. Whenever one of the puzzle pieces latches on in the right place, the marble has climbed down into a deeper valley—the particles have found a more stable configuration. Innumerable such descents later, nature has created one of those wonders where trillions of atoms or molecules are arranged in a perfect geometric pattern—all without a guiding hand.

Molecules that explore vast energy landscapes are behind much of the inanimate world's beauty, from the atmospheric wonders of snowflakes, to crystalline rocks like granite, and to gemstones like diamonds, rubies, and emeralds. And just as their materials are diverse, so too are the amount of heat and the rate of cooling needed to build them. To get a bucky-ball to self-assemble, carbon must be heated to thousands of degrees, far above the temperatures at which a snowflake grows, but then it takes mere milliseconds of cooling to create that perfect ball.[14] In contrast, to solve the complex puzzle needed to assemble a large diamond, nature often needs more than a billion years.[15]

Most crystals in nature do not have the perfect shape dictated by the arrangement of their atoms with the lowest potential energy.[16] The bewildering diversity of snowflakes makes that plain. Their shapes are often very different from the ideal shape of crystalline water ice—a hexagonal prism. To be sure, the smallest snow crystals often do display this perfection, but larger ones don't. They grow from such a prism, a tiny crystalline seed inside a swirling cloud of ice-cold water vapor, and as they do, fewer water molecules tend to attach along the prism's flat surfaces than attach near the prism's protruding edges—they get caught there as they drift through the air. In other words, a snowflake grows more slowly in some places and faster in others. Where it grows faster, branches can sprout in a process that physicists call a branching instability. These branches can father new branches, and so on. This is how the familiar arborescent filigree of a full-grown snowflake builds itself.[17]

When, on a cold winter day, we watch millions of these creations quietly rain from the sky as far as the eye can see, we can begin to grasp the vastness of nature's creative landscapes. Each snowflake is far from an amorphous jumble of water molecules. It is a good but not perfect solution to the problem of minimizing potential energy. Each one occupies a different deep—but not the deepest—valley in the energy landscape of water ice. And that's why each one is unique.

Snowflakes and other crystals not only teach us that imperfection can harbor great beauty, but they also make bucky-balls all the more remarkable. That's because the landscape of bucky-balls also has valleys beyond measure. Some are shallow, with an amorphous jumble of carbon atoms. Others are deep, but not quite as deep as the deepest valley of that perfect carbon soccer ball—they correspond to imperfect ovoids with various distortions of the perfect soccer ball.[18] Yet, when the conditions of the experiment are just right, the majority of all carbon atoms aggregate into bucky-balls. Every single one of them has found the deepest valley.[19] The lesson: the right amount of vibration can be so powerful that it conquers even the most complex landscape.

It is no coincidence that we encountered similar jostling in the inverted landscapes of life's evolution. Many of these landscapes cannot be conquered by selection's steady uphill moves, because their topography is no less varied than the energy landscapes of the inorganic world. The jostling in evolution's adaptive landscapes does not come from heat, of course, but from heat's analog, the tremors of genetic drift, which allow small populations to escape shallow adaptive peaks. When these tremors are very strong—in the smallest populations—they can drive a population across the landscape no matter how rough the terrain. The smaller a population is, the stronger genetic drift becomes, the more violent these tremors are, and the faster the population explores its adaptive landscape. Drift prevents a population's capture by a shallow peak, just like heat helps a molecule, crystal, or atom cluster avoid entrapment in a shallow valley of an energy landscape.

Heat is as important to creating the beauty of the inorganic world as genetic drift is to the beauty of the living world. Superficially, heat and drift are completely unrelated, but deep down, they are means to the same end—conquering the landscapes of creation. Both of them permit imperfection to find perfection.

Incredibly, the beauty of diamonds and butterflies comes from the same source, even though they are so different. Perhaps, then, this source is important far beyond nature. Some scientists and engineers have had this suspicion for decades. More than that, they found proof, which helps them not only find perfect solutions to the hard problems they face, but also allows them to do something even more important: delegate the task of solving such problems. Not to people but to computers—creative computers.

Chapter 6

Creative Machines

When we say that trucking is tough business, we usually think of the drivers and their lonely hours on the road, extended separations from family, exercise-deprived work days, and artery-clogging food eaten at highway rest stops. But for the companies employing them, life is no bed of roses either. Elbowing their way through a highly competitive market with razor-thin profit margins,[1] the name of their game is efficiency—in fuel use, downtime, and, most of all, routing. The difference between success and failure can hinge on a few percent of excessive mileage.

Routing a truck seems easy enough. Load goods for delivery at depot, mark customers on map, connect the dots, and off you go. But don't be fooled. A truck has limited capacity, needs to deliver within a certain time window, may have more than one depot to visit, and must visit multiple drop-off points. Even the problem of finding the shortest route to these drop-off points would stump any driver, rare geniuses aside. Just consider the numbers. Six different routes are possible to visit three different customers—ok, your average driver might still be able to figure

that one out. But with twenty-four different routes for four customers, and 120 routes with five customers, the problem gets
harder. Ten customers? Three million routes. Fifteen? More than
a trillion. The numbers rise so fast it is scary, and they become
large beyond comprehension for realistic numbers of hundreds of
customers.[2] And that's just for one truck. Companies like Fed-
eral Express have more than forty thousand trucks, not to speak
of their six hundred planes that fly to more than two hundred
countries. Figuring out the best route takes more than a genius.

The problem—to deliver the largest amount of cargo at the
lowest possible cost—is a mathematical one. Because it is complex, and because companies need to solve it every day, it's a
problem ready-made for computers. But the answers are not, unless you have an algorithm—a prescribed sequence of simpler
computations that computers can follow—that is able to solve
the problem day after day after day.

Finding algorithms is another tough business—not that of
truckers, but of computer scientists. The problem of finding the
shortest delivery route is so famous (and hard) that computer scientists have created their own acronym for it: the VRP, or vehicle routing problem. It is a close cousin of an even more famous
and older problem that goes back to the nineteenth century, the
traveling salesman problem, or TSP. This problem arises when a
salesman needs to visit multiple customers and tries to keep his
route as short as possible. The problem was not only important
for the up to three hundred and fifty thousand traveling sales-
men who peddled their wares throughout the United States in
the late nineteenth and early twentieth centuries. It also came
up in other professions, such as traveling preachers and traveling judges. Judges, for example, journeyed through their districts

on a fixed route of towns—a circuit—that held court on specific days of the year. The name of *circuit courts* for US regional courts survives this long-extinct practice.[3]

Thousands of computer scientists have grappled with difficult problems like the TSP and VRP, and not because they care so deeply about truckers and salespeople, but because these problems pop up in many other areas. Take chemistry. To map the atoms in a complex molecule, a chemist can crystallize the molecule, shine an X-ray beam onto the crystal, and measure how its atoms diffract the beam. The problem is that she needs to measure the beam's diffraction not just from one angle, but from hundreds of different angles. In other words, she needs to rotate the crystal through hundreds of positions. The faster she can do that—the shorter the path between these positions—the shorter her experiment will be.

An astronomer who wants to observe hundreds of stars or galaxies in the night sky faces a similar problem. For each observation, a telescope must be rotated into a precise position, a job that computer-driven motors handle for large telescopes. Modern telescopes are hugely expensive and are shared by researchers all around the world, who must line up to obtain precious observation time. The faster a telescope can visit all desired positions—the shorter the path through them—the more time is available for observations, and the more efficiently the telescope is used.

Likewise, when computer-chip designers lay out millions of transistors on a new chip, they need to keep the wiring between transistors short. Otherwise, their design would waste precious square millimeters of the chip's real estate. Plus, a chip whose electrons have to travel long distances between transistors would squander costly energy.

Finding the shortest path of rotating a crystal, the most economical way of positioning a telescope, the shortest wiring route between transistors—these are all different flavors of the same mathematical problem.[4]

But even more important than these applications of the routing problem is that the problem harbors profound lessons about the very nature of hard problems everywhere—the kind that require the most creative solutions. It is another stepping-stone to understanding creativity in general, because solving it requires the same tricks that nature has used to create bucky-balls and buttercups. They are the tricks needed to traverse landscapes vast beyond comprehension.

Take the diagram in Figure 6.1, which stands for a route that a delivery truck takes from a depot to visit each one of ten customers. The customers are arbitrarily labeled with the numbers one through ten. Underneath the diagram, a string of numbers stands for the order in which each customer is visited. This string encodes the route itself.

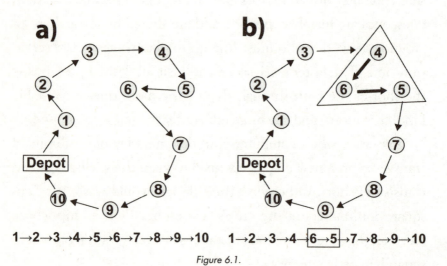

Figure 6.1.

There are more than three million possible different routes
to a mere ten customers, and each route can be written as one
string like this. Each of these routes has a cost, expressed in distance traveled, time expended, fuel combusted, or—for the envi-
ronmentally conscious—carbon dioxide exhaled by the engine.[5]
Just like every conceivable DNA string of a gene encodes a
solution—good or bad—to one of life's many problems, every sequence of customers to be visited encodes a solution to the routing problem. Most of these solutions are poor, excessively long
routes, but a few of them are good, short, efficient routes. And
just like a DNA sequence—a genotype—specifies a location in
an adaptive landscape, each route corresponds to a place in yet
another landscape, that of all solutions to the routing problem.
The elevation in that location corresponds to the length of the
route. It is the quality of the solution, the analogue to an organism's fitness.

This cost landscape is yet another abstract and high-
dimensional landscape, not the three-dimensional landscape in
which trucks deliver goods. The landscape's profile of undulating
dunes and cragged mountains, shallow dips and deep ravines, is
the cost profile of different routes. This cost profile differs from
evolution's landscapes in the same way as the energy landscapes
of atoms and molecules do: its peaks are the worst places to be
because that's where the longest routes dwell.

Any algorithm to route vehicles or salespeople is a prescription to scour the cost landscape for the deepest valleys—those
corresponding to the best and shortest routes. The algorithm is a
bit like a molecule seeking an especially stable configuration of
atoms, a deep valley in its energy landscape. It explores the cost
landscape of the routing problem, just like vibrating molecules

explore their energy landscape, and just like evolving organisms explore their adaptive landscape.

In evolution's landscapes, the smallest steps correspond to point mutations, those DNA mutations that cause single-letter changes. Unfortunately, these same kinds of steps don't work in the cost landscape of the routing problem. Imagine you replace a single customer, say customer five, in the sequence underneath Figure 6.1a, with an arbitrary other customer, say customer ten. That would be the closest analogue to how point mutations change DNA. But it creates two problems. First, because you have eliminated customer five, the resulting route would not visit this customer at all. Second, the route would visit customer ten twice. In other words, mimicking point mutation violates two basic rules for vehicle routing: all customers must be visited, and each customer can be visited only once. Neighboring routes cannot just differ in a single customer. We need a different kind of "mutation."

What we need is to permute or swap the order in which two customers are visited, as Figure 6.1b illustrates for customers five and six. Swapping ensures that each customer is visited exactly once. More than that, any route connecting any number of customers can be created with the right sequence of such swaps. In other words, starting anywhere on this landscape, you can travel anywhere else by customer swaps, which are the smallest possible steps on this landscape.[6]

Such travel ultimately has one goal: to find the deepest valley. Computer scientists have invented many algorithms—some better, some worse—to meet this goal. Let's look at an especially simple one.

Choose an arbitrary route—no matter how long and inefficient—and change it through a random swap of two customers. If the swap reduces the length of the route, keep it. If it does not, go back to the starting point. Then repeat. Swap another two randomly chosen customers, keeping the new route only if it is shorter. Keep repeating. Each successful swap changes your place in the solution landscape. Step by step—swap by successful swap—you will eventually reach a route that cannot be shortened further by a swap.

This algorithm will always descend in the landscape, aiming to find the lowest point. It is the analogue to natural selection's uphill movement in a fitness landscape. Where natural selection permits only steps that *increase* fitness, the algorithm permits only steps that *decrease* a route's length. If the starting solution was a marble on that landscape, the algorithm would act like gravity on the marble, pulling it downward toward the nearest resting point.

Algorithms like this are important because they are simple. Computer scientists also call them "greedy." And their greed need not be bad. If the cost landscape is like a meteor crater weathered and smoothed by the ages, then any sequence of downhill steps will get you to the bottom.[7]

But that's a big *if*.

Vehicle routing belongs to a huge class of problems that computer scientists call combinatorial optimization problems. That's because each solution consists of various building blocks—customers, in this case—and searching for a best or optimal solution requires combining these building blocks in different ways. When routing trucks, you combine the same customers in different orders to find the shortest route. When designing an electric power grid, you combine multiple power plants in different locations to ensure that customers get enough electricity. When scheduling nurses in a hospital, you assign different

combinations of nurses to different shifts. When planning a battle, you maximize your enemy's hurt by attacking different tar-
gets with different combinations of weapons. And so on.

Not all combinatorial optimization problems are hard, but all
of them have a landscape of solutions. And after exploring and
mapping and studying these landscapes for over half a century,
computer scientists have learned a key lesson about what makes
a problem—*any* kind of problem—hard.[8]

A problem is not hard simply because it has many possible
solutions—all combinatorial optimization problems do. It is
hard for a reason already familiar from looking at self-assembling
molecules and adaptive landscapes: the solutions form a rugged
moonscape, with myriad shallow craters, fewer deep craters, and
no more than a handful of deepest craters, which can be nearly
impossible to find among all the others.

The cost landscape of vehicle rerouting is just like that, with
countless craters big and small. Chances are that an algorithm
that can only descend will climb down a small crater—there are
so many of them—and quickly get stuck at a bad solution from
which there is no escape. Even if by dumb luck it descended into
a large crater, it might not get very far. That's because big craters
are pockmarked by smaller craters, which themselves contain
even smaller craters. The algorithm might end up in a tiny crater
close to the rim of the big one, with a route that's far longer than
it could be and no way to escape.

Hard problems have another nasty property: they get even
harder—much harder—as they grow in size. As the number of
customers, power plants, nurses, or weapons increases, the number
of craters—the technical term is *local minima*—increases exponentially. And just like those minima in a molecule's energy landscape,

their numbers quickly overwhelm even the most powerful computers. Whereas the vehicle-routing landscape for ten customers has some one hundred local minima, the same landscape for fifteen customers can harbor more than one thousand local minima,[9] and for realistic numbers—a hundred or more customers—the number of minima is too large to count. Only one of them is the best route, or *global minimum*, and finding it for large problems is like finding a drop of oil in an ocean. In practice, the best one can hope for is a reasonably good solution, one of multiple local minima that are deep, but not quite as deep as the global one. They are the snowflakes of difficult problem solving.

Any algorithm scouring a landscape like this for a good solution faces an already familiar problem: to get out of a shallow valley and explore a deeper one, it needs to overcome the relentless downward pull of gravity. It's the same kind of problem evolution faces with its own greedy algorithm—natural selection.

Fortunately, where greedy algorithms fail, others can succeed. One such algorithm is called *simulated annealing*.

The word *annealing* comes from metallurgy and refers to a treatment that renders steel and other metals less brittle and more workable. Bladesmiths, for example, use it when forging swords. When a piece of steel is annealed, it is first heated to temperatures at which the vibrations of its iron atoms become so powerful that these atoms begin to change their location and drift through the metal. The piece is then cooled down very slowly, which allows atoms to form small crystals and renders the steel more malleable. The treatment's central ingredient is heat—the same that allows molecules to explore their energy landscapes and discover buckyballs, gold clusters, and diamonds. And that's not a coincidence. It reflects the deep connections between different realms of creation.

Simulated annealing—the name says it all—simulates the process of annealing with a computer algorithm. The algorithm explores the solution landscape of a problem and journeys the landscape step-by-step. (If the problem is how to route trucks, each step might swap two customers.) After each step, the algorithm computes whether the step leads to a better solution, such as a shorter route. If so, it accepts the step. This much is identical to a greedy algorithm. But the crucial difference is this: even if the step leads to a worse solution, the algorithm accepts the step with some probability. Early in the algorithm's journey, this probability is high—the algorithm might be just as likely to accept uphill steps as downhill steps. As time goes on, while the algorithm explores the landscape, "cooling" begins—the algorithm steadily lowers the probability of accepting uphill steps. After a thousand steps, the algorithm might accept only half of all uphill steps; after ten thousand steps, it might accept only one in ten; after one hundred thousand steps, only one in one hundred, and so on, until eventually it accepts only downhill steps.

To see how this procedure resembles the cooling of a material, compare the energy landscape of a bucky-ball or snowflake from Chapter 5 with the solution landscape of a problem. If a greedy algorithm is like gravity pulling down a metaphorical marble—a configuration of carbon atoms or a solution to the routing problem—then simulated heat acts like the jitters that allow the marble to escape that pull. Just like the strong jitters of particles in a hot material allow uphill motion through an energy landscape (to more unstable carbon molecules), so too does the early phase of simulated annealing allow moves to much worse solutions (to much longer routes). Whenever that marble lands in a valley early on, the jitters will quickly jerk it out of there, no matter how deep the valley—how stable the molecule or how good the solution. During this early stage, the marble can explore wide swaths of the landscape, however rough the terrain. As cooling progresses, the jitters in the landscape become weaker. The marble is more likely to move downhill, toward more-stable molecules or shorter routes, and become locked into the current valley. Escape is especially unlikely if that valley is deep, because it would require many consecutive uphill steps. But the deepest valleys in a rugged landscape often contain shallow valleys, and even if the landscape's jitters have become weak, the marble can still climb out of a shallow valley and roll into the next one. Only at the very end will the marble become confined to a valley. It will travel to the bottom of this valley, to its final resting place, which corresponds to the most stable molecule that nature can discover or to the best solution that an algorithm can find.

The appeal of simulated annealing lies in a simple mathematical fact: one can prove that the deepest valley—the global minimum or shortest delivery route—can be found with certainty if cooling is slow enough; that is, if the tremors' intensity declines slowly enough.[10] These are the same kinds of tremors that enable genetic drift to shake up populations that evolve on an adaptive landscape. Evolution in a small population, where drift is strong, is like the initial stage of simulated annealing. And simulated annealing itself is a bit like evolution in a population that starts out small and grows, such that genetic drift becomes weaker and weaker over time.[11]

Simulated annealing, however, has one advantage over biological evolution. Because it is a computer algorithm, we can control every detail about it. In contrast, no mastermind of biological evolution manipulates evolving populations for its maximum benefit. Population sizes, for example, depend on many environmental vagaries, including climate, food, and competitors, whereas temperature, which is crucial for the success of simulated annealing, can be controlled with great accuracy via a computer algorithm. Because of this ability, simulated annealing and other human algorithms may be able not only to solve hard problems, but to solve them better than even nature could.

But couldn't we apply the same idea to evolution? Couldn't we make evolution more controllable by simulating it inside a computer? This is not so far-fetched, because Darwinian evolution is itself a kind of algorithm, a sequence of simpler steps: mutate, select, repeat. Since this algorithm transforms not bits and bytes but living matter, it could be applied to simulated DNA, simulated phenotypes, and simulated organisms. And in such simulations, we could control every little detail—how often mutations occur, how rigorously we select the best variants, whether we allow sex and recombination, and so on. This level of control could again help us improve the algorithm's problem-solving ability. What is more, engineers could not only modify the algorithm, but also exploit it to solve problems they care about, problems that may be as alien to nature as routing trucks.

The dream of harnessing the algorithms of evolution for our benefit is not new. It goes back at least to computer science pioneer Alan Turing. In a seminal 1950 article entitled "Computing Machinery and Intelligence," Turing envisioned machines that do not just think but also learn, and he argued that a "random element" analogous to DNA mutation may help them do so.[12]

The realization of Turing's vision had to wait until the 1960s and 1970s, when computers became powerful enough to simulate evolution. That's when a new research field called evolutionary computation emerged. One of its pioneers was John Holland, an engineering and computer science professor at the University of Michigan. Gifted with an ebullient and cheerful personality, Holland exuded joy about science from every pore of his body. His enthusiasm for new ideas was contagious, and his contempt of established dogma profound. These traits helped him become a pioneer of evolutionary computation in the 1970s. That's when he developed a class of computer algorithms that mimicked evolution—he called them genetic algorithms—and remain among the most powerful in evolutionary computation.[13] Through his work and that of others, evolutionary computation ballooned into a new discipline of science that today employs thousands of researchers, whose goal is to improve on nature's algorithmic handiwork. And improve it they have.

Unlike simulated annealing, which drives a single marble across the solution landscape of a problem, a genetic algorithm uses multiple marbles, an entire population of them. Inside a computer, each member of this population contains a "chromosome," a string like that in Figure 6.1a that encodes a solution—good or poor—much like DNA encodes a well-adapted or ill-adapted organism. Each string can "mutate" to a new one, for example, by swapping two random customers in a chromosome encoding a solution to vehicle routing. The computer then chooses the best solutions in the population to select a new and improved population. Mutate, select, repeat—just like biological evolution.

Because genetic algorithms follow biological evolution's time-honored principles, they also face the same hurdle in solving tough problems: the unforgiving nature of selection. Much like selection in the jungles of Borneo, selection inside a silicon chip is a poor guide to finding the highest adaptive peaks or the lowest-cost valleys on a rugged landscape. Selection is shortsighted, so it exterminates imperfect solutions. Even worse, it prohibits the kinds of failures that can be essential for eventual success.

Fortunately, genetic drift can help genetic algorithms avoid this trap, just as it does for biological evolution. Most genetic algorithms evolve populations that comprise anywhere from ten to a few thousand individuals—simulating much larger populations simply requires too much computer memory and time.[14] These populations are much smaller than the billion-strong populations of microbes and are more like those of large animals and plants, where drift is strong enough to overpower some of selection's aversion to downhill steps. In other words, genetic algorithms overcome the trap of relentless selection for free, courtesy of their modest populations, themselves a consequence of limited computing power.

Genetic algorithms also simulate sex. In their populations, some individuals "mate" and produce "children"—new solutions to a problem—by swapping chunks of their simulated chromosomes, just like recombination swaps chunks of real DNA chromosomes inside organisms.[15] This artificial recombination can also teleport individuals through a solution landscape, exploring new solutions that would take forever to reach with smaller steps. And in a development that mirrors the spectacular success of sex in biological evolution, recombination turns out to be so important in genetic algorithms that some scientists program the algorithms to have recombination cause 90 percent of all chromosome changes, leaving only the remaining 10 percent to mutations.[16]

Genetic algorithms and simulated annealing are only two animals in a large zoo of problem-solving algorithms. That zoo houses many other creatures with exotic names, like branch-and-bound, linear programming, and tabu search. They were created by scientists who pored over old, important, and difficult problems like that of the traveling salesman and accumulated deep knowledge about their mathematical structure. This knowledge helped them illuminate different shortcuts through the solution landscape and design algorithms to excel at finding these shortcuts. And these algorithms can perform impressively on hard problems, like that of finding shortest routes.[17]

For example, the shortest route connecting 666 of the world's tourist destinations has been known since 1987. That's when an algorithm succeeded in finding it among more than 10^{1500} possible routes. In 1998, after another decade of algorithm design and improvements in computing power, researchers at Rice University found the shortest route connecting 13,509 cities and towns in the United States.

An even bigger problem would interest Santa Claus: find the shortest route between all 1.9 million known human settlements. Even though this problem is still unsolved, the algorithm that discovered the best *known* route—7.5 million kilometers long— has done a great job: one can prove mathematically that this route is at most 0.5 percent longer than the unknown shortest route. If Santa ran a trucking company and had to worry about profit margins, he could rest easy.[18]

That's what state-of-the art algorithms can accomplish when wielded by teams of computer scientists with huge amounts of computing power. They can sometimes find the best solutions even for problems of staggering complexity. And when they cannot, it's because even their shortcuts are no match for the complexity of a solution landscape. For example, even though simulated annealing is guaranteed to find the deepest valley in any landscape— eventually—some problems require cooling so glacially slow that it might take thousands of years to find the deepest valley.

Most state-of-the-art algorithms are made for problems that are so well studied that their solution landscape is visible in hazy outline. But many of the problems that engineers face day-to-day—how to reduce the emissions of a combustion engine, increase the efficiency of a solar panel, or decrease the side effects of a drug—are one-of-a-kind, novel problems where the landscape lies in the dark. In these situations, we are blind, and our algorithms need to explore solution landscapes blindly. That's where genetic algorithms shine, because they explore this landscape just like biological evolution explores adaptive landscapes, through blind mutation and selection. If that blindness seems like a fatal handicap, just think about the marvels that biological evolution has created.

Today's algorithms produce remarkable feats, but most people would not call their products—a shortest delivery route, optimal staffing at a hospital, or a humming electric grid—creative. To see what's missing, it is useful to take another look at genetic algorithms, because they follow the same steps that evolution has used to create living beings as spectacular as redwood trees and blue whales. We have no trouble seeing these species as the results of a creative process, so what's the difference? It cannot lie in *how* an algorithm creates something new, because the *how* is similar between biological and simulated evolution. It must lie, then, in *what* the algorithm creates—in its chromosomes and what they encode.

Whereas the chromosomes of most genetic algorithms are strings of numbers that represent abstract concepts like the order in which customers are visited, the DNA of a real chromosome is a string of molecules that encodes objects in the real world, from minute proteins to huge dinosaurs. Whenever evolution creates a protein to digest a new food, sense a new smell, or disarm a new antibiotic, and whenever it creates a gazelle that runs a bit faster, a bird that soars a bit higher, or a fish that dives a bit deeper, it solves an immense combinatorial optimization problem. The elements of this problem are the same old four DNA letters. Evolution combines them in ever-new ways, scouring an adaptive landscape to find its highest peaks.

The message could not be clearer: for true creation, choose the right building blocks.[19]

Among the first to understand this principle—and implement it—was John Koza, a former Ph.D. student of genetic algorithm pioneer John Holland. Beginning in the 1990s, Koza and his collaborators designed genetic algorithms whose chromosomes encoded the components of electric circuits, such as resistors and capacitors, as well as how these components were wired together.[20] They used these algorithms to evolve not only new kinds of circuitry, but circuitry that met a rigorous legal standard for creativity. This standard is not so different from how psychologists define creativity—an original solution to a problem—except that it clearly defines the meaning of *original*: worthy of receiving a patent. Every invention solves some problem, but only the most original ones are patentable. In the words of US patent law, an invention is patentable if it is not obvious to a person with ordinary skills in the invention's field. Koza's genetic algorithms helped create electronic circuits that met this standard.

In the course of a few years, these algorithms created wiring diagrams for more than a dozen functioning circuits, such as the kind of low-pass filter that sends rumbling low-frequency sounds to a sound system's subwoofer. Several of these circuits had been invented by human engineers and patented by companies such as AT&T and Bell Labs years before Koza's work. But their reinvention made a powerful point: algorithms can be creative, and their products can rival those of human creators.

Another case in point: in 2005, the algorithms of Koza's team discovered a new kind of controller—the kind that holds the speed of your car steady when it is on cruise control. The controller was sufficiently innovative that the team received a patent for it.

It was the first patented invention made by a machine.[21]

The creative potential of evolutionary algorithms does not end there. With chromosomes that encode the surface texture of solar cells, a genetic algorithm can evolve surfaces that excel at trapping light and increase solar-cell efficiency. In this light-harvesting problem, many suboptimal surfaces—shallow valleys—exist, and mechanisms like recombination are needed to leap over these inferior solutions.[22]

With chromosomes that encode the complex lens system of telescopes or binoculars, a genetic algorithm can evolve a wide-field eye piece superior to one patented by human lens designers.[23] And with chromosomes tuned to the parts of an antenna, a genetic algorithm can equip NASA's space missions with new kinds of antennae that can operate in outer space. Such an antenna was placed onto satellites in one of NASA's space technology missions in 2006. It bears the distinction of being the first human hardware in space that was not designed by humans but rather was evolved by an algorithm.[24]

And as for engineering, so too for science. We admire the ingenuity of scientists like Newton and Galileo who can distill mathematical laws of nature from observing a physical system like a planet or a pendulum. As it turns out, algorithms mimicking evolution can do the same thing. In a 2009 study, scientists at Cornell University described an algorithm that can discover the equations of motion of a simple pendulum—one weight suspended from a pivot—or a more complicated double pendulum—two weights and two pivots. All the algorithm needs is data on the pendulum's motion as well as the building blocks of the solution. These building blocks are variables and mathematical functions. In other words, such an algorithm does not manipulate transistors, lenses, or wires, but it mutates and recombines mathematical functions until it has found a combination of these functions that describes the data perfectly.[25]

Given such feats, we should not be surprised that genetic algorithms have been called "Thomas Edison in a box."[26] Or should we say, "Galileo in a box?" The age of creative machines clearly has arrived.

Many people recoil at this thought. It is an inconvenient truth on a par with the realization that the earth is not the center of the universe, that we share a common ancestor with chimpanzees, and that we are not the only animals to recognize ourselves in a mirror. Discoveries like this have chipped away at our self-importance over the centuries since the Copernican revolution. But, like it or not, they are here to stay, just like the discovery that we humans have no monopoly on creativity.

And here is a thought that may soften the blow: the word *machine* still evokes in many of us visions of the products of a bygone era—eighteenth-century contraptions like steam engines, mechanized looms, or chronometers made of springs, rods, bearings, chains, and gears. But creative machines are not like this. Not at all. Nor are the algorithms they execute. An algorithm is often compared to a recipe followed slavishly by a chef—the machine—but no such algorithm could produce anything new. Every time you execute such an algorithm, you get the same dish. A genetic algorithm—and algorithms like simulated annealing—are different from a conventional recipe because they are stochastic. This means that randomness is essential to them, just as in biological evolution, where DNA mutations and recombination occur at random locations in a genome and where every such random change can alter evolution's path. The products of creative machines can therefore be just as unpredictable, original, and unique as the products of biological evolution—or as human works of art.[27]

Art created by computers may feel even more disturbing than technology created by them because the uniqueness of great art feels at odds with the notion of an algorithmic search through a landscape. To resolve that dissonance, it helps to recall once again that the landscapes we have encountered are vast enough to admit innumerable creations that are as unique as snowflakes. Because an algorithm's journey through such a landscape is unpredictable, it may discover each such creation only once.

A striking work of art in the Ryoanji Zen temple of Kyoto illustrates the kind of problem that an artsy algorithm might solve. This temple houses a famous rock garden with a simple design consisting of fifteen carefully arranged rocks on a bed of gravel. The garden exudes an atmosphere of uncanny harmony and serenity. A UNESCO World Heritage Site, it is visited by thousands of tourists every year. Sadly, we have no idea how the artist created the garden's powerful effect because the garden is some five hundred years old and the artist left no record. However, cognitive scientist Gert van Tonder and two collaborators used mathematical tools to analyze its design and found a surprisingly simple pattern. From the temple's veranda—the place from which the garden is best viewed—the symmetry of the rock arrangement evokes the fractal pattern of a branching tree. It's a kind of pattern that humans find attractive, perhaps because of our evolutionary roots in African savannas. In other words, part of the rock garden's appeal lies in an abstract image extracted from a motif omnipresent in nature. Consciously or not, the artist solved the problem of creating a serene ambience by embedding this image in the garden's rock arrangement.[28]

Understanding the power of a work of art is one thing. Creating it is another. But even there algorithms are making great strides. And it is no coincidence that some of their most promising achievements have occurred in creating musical compositions. It's because music provides the ideal building blocks for a creative algorithm.

The elementary building blocks of musical compositions are motifs, snippets of melodies that can be varied in ways that composers and improvisers have known and classified for centuries. Composers "augment" a motif by slowing it down, "diminish" it by speeding it up, invert it, permute its notes, play it backward, transpose it into a new key, and so on. That such "mutated" motifs could be randomly combined into novel pieces of music was already known in the eighteenth century, when musical dice games, or *Musikalische Würfelspiele*, as they are known in German, became popular parlor games in Europe. In such games—the best-known has been attributed to Mozart himself—dice are rolled and cards are drawn to select small snippets of music from a larger collection, which are then concatenated to compose a waltz, a polonaise, or a minuet.

The computer age took this tradition to new heights.[29] It brought forth a branch of artificial intelligence research that uses algorithms to create musical compositions. One of its pioneers was David Cope, an accomplished American composer whose own work was lauded by critics and performed at Carnegie Hall when he was still a young man in the 1980s. During this time, while trying to write a commissioned opera, he experienced a severe bout of writer's block that lasted for months. In his despair to keep producing the music that fed his family, he started to ask whether computers could help him compose. His efforts eventually culminated in a program he called Experiments in Musical Intelligence, or, more affectionately, "Emmy."[30]

When Emmy was fed a composer's work, she—yes, that's how Cope refers to his program—analyzed the work to find signature elements of the composer's style. She then varied these building blocks—mutation—and created new combinations of them. The results were novel compositions in the style of a Bach, Mahler, or Vivaldi. And they poured forth at breathtaking speed. In Cope's own words: "You pushed the button and out came hundreds and thousands of sonatas."[31]

We value any one composer's work partly because there is a limited amount of it. After all, composers are mortal. Cope realized that Emmy's unlimited creativity would diminish the value of any one composition. And so he let Emmy die. After several years, he mothballed her digital brain, but not until he had published more than a thousand of her compositions.[32] Some of them appeared on CD, and others can be admired on internet platforms like YouTube. Some listeners deride these compositions as soulless, while others find them deeply touching. But many people—unless they are told beforehand—would not identify them as machine-made, making Emmy's oeuvre a musical version of Alan Turing's famous test for distinguishing artificial intelligence from human intelligence. In Turing's original version from the 1950s, the test needs at least three participants: a human, an artificial intelligence, and a judge. The judge poses questions to the other two participants, but he cannot see them, and he receives only written answers to his questions. He thus needs to guess whether each answer comes from the human or the artificial intelligence. If he guesses wrong, the artificial intelligence has passed the test.

A similar test was performed with one of Emmy's Bach-style compositions at the University of Oregon. An audience—the judges—listened to Emmy's composition, a composition by Bach, and a third, Bach-style composition written by a contemporary composer, all of them performed by a professional pianist. When the audience members were asked to tell which was which, they got it wrong. They declared that Emmy's composition was a genuine Bach and that the other human composition was written by the computer.[33] Emmy had passed this musical Turing test.

Some critics nonetheless snivel at Emmy's compositions and call them "imitations." To them, Cope responds that "all composers quote and allude to music, including their own" and "I can find passage after passage in Mozart's symphonies that quote and paraphrase similar passages in Haydn. However, we still revere both composers."[34] Cope has viewed music composition as "inspired plagiarism," and when he calls Emmy's compositions "recombinant music" it is not to diminish their originality.[35] Rather, it is an acknowledgment that composers often do not create from scratch. Instead, they reuse, modify, and recombine what has come before them. Like nature does.

Even though mutation and recombination played a role in Emmy's creations, she did not execute a genetic algorithm because she did not evolve a composition by continual mutation and selection over multiple generations. But other algorithms do. One of them is Melomics, whose name comes from "melody" and "genomics." In 2012, Melomics created a composition that was performed and recorded by the London Symphony Orchestra and its top-notch professional musicians. Some listeners have described the recording as artistic, delightful, and expressive, but as with most modern music, opinions are divided.[36]

Algorithms can also create in other genres, like jazz. One such algorithm, called the Continuator, is the brainchild of François Pachet from the Sony Computer Science Laboratory in Paris. The Continuator "listens" to a human player's improvisation and learns to improvise in the same style. Pachet subjected these improvisations to a Turing test, in which the professional jazz pianist Albert van Veenendaal improvised on the piano. Pachet employed not one but two judges, both of them music critics, who had to determine whether an improvisation they heard came from van Veenendaal or from the Continuator. They could not.[37]

The following story is fodder for yet another Turing test:

> *Friona fell 10–8 to Boys Ranch in five innings on Monday at Friona despite racking up seven hits and eight runs. Friona was led by a flawless day at the dish by Hunter Sundre, who went 2–2 against Boys Ranch pitching. Sundre singled in the third inning and tripled in the fourth inning . . . Friona piled up the steals, swiping eight bags in all.*[38]

The story summarizes a Little League baseball game. That's unusual—Little League does not receive much press coverage— but even more unusual is this: the story was written by a computer. It is the product of the Chicago-based company *Narrative*

Science. The algorithms of this and other companies prepare reports on sports events, corporate earnings, or real estate transactions, created instantly from available data when ordered by one of the company's media clients. Many of these clients prefer anonymity, what with journalists already worried about their jobs, but they include well-respected outlets like *Forbes* magazine or the Associated Press, which uses them to produce three thousand financial reports per quarter.[39] Chances are good that a magazine you have been reading—online or off—contains some automated reporting.[40] What is more, a 2014 study showed that people cannot tell such texts from human texts.[41]

Computers are beginning to invade multiple domains of human creativity. They haven't yet produced a Beethoven, Beckett, or Picasso, but their products are already more than curiosities. They appeal to human audiences and are going commercial, in companies like *Narrative Science* or *Jukedeck*—a British start-up selling algorithmic music on demand for video soundtracks.[42] What is more, if other branches of artificial intelligence are any guide, their creativity will be improving faster than we think. Just remember how explosively the power of chess computers increased: they beat a world champion like Gary Kasparov less than fifty years after their humble beginnings.

Regardless of how long creative algorithms may take to exceed human performance in any one domain, they have already taught us two important lessons. The first is that the right building blocks are crucial to producing creative works. These building blocks are obvious in domains like electronics or music, but less obvious in domains like the visual arts. The second lesson is that creativity requires the conquest of landscapes as complex as the adaptive landscape of evolution, the cost landscape of vehicle routing, and the energy landscape of bucky-balls. In these land-scapes creativity must avoid the fatal attraction of nearby hills or shallow valleys, which harbor the obvious and mediocre. They are death traps on the path toward the original and superior.

The universe has managed to sidestep these hills since even before life arose—in creations as different as bucky-balls forged in a sun and snowflakes drifting down from the sky—all thanks to the power of thermal vibrations. Life has added its own tricks—genetic drift, recombination—and computer scientists have added yet others, all in the service of finding the highest peaks or lowest valleys in a landscape of creation.

The books of nature are still easier to read than those of our brains, and that's why we have not yet mapped the landscapes that our minds explore. What we can safely say, however, is that landscapes must be as central to human creativity as they are ev-erywhere else. That's because creativity is about solving difficult problems, and—a key insight of twentieth-century computer science—a problem's difficulty is encapsulated in its solution landscape. No matter whether a problem is solved by an evolv-ing organism, a computer, or a human mind, a landscape of solu-tions needs to be conquered. What is more, the landscapes of our minds must have many peaks and valleys—mediocre stories, corny poems, and lousy compositions—otherwise anybody could write like Tolstoy or compose like Mozart. But before we explore how human minds avoid the trap of mediocrity, it will be useful to learn that Darwinian evolution and the mind's creative pro-cess are more similar than one might think.

Chapter 7

Darwin in the Mind

On April 26, 1937, the tiny Basque town of Guernica was overflowing with people attending its Monday market, when a squadron of German bombers and fighter planes attacked. Three hours, forty tons of bombs, and thousands of machine-gun bullets later, hundreds of unarmed villagers were dead.

Words fail to capture the horror of Guernica, but art might be able to. If any painting can capture human anguish, *Guernica*, Pablo Picasso's contribution to the 1937 Paris World Fair, is the one. A woman wailing over her lifeless child, a dismembered warrior with a broken sword, a terrified figure engulfed in flames, and a screaming horse with a gaping wound in its flank suffer amid a jumble of broken bodies. *Guernica* is a stark testimonial to suffering and pain.[1]

Guernica not only reveals the horrors of war, it also reveals a great deal about the creative mind. That's because Picasso dated and numbered the forty-five different sketches he drew to prepare the painting. Some of these sketches explore different arrangements of people, animals, and body parts, others show variants of the mother with the dead child, the head of the dead warrior, and the gored horse. Some sketches are easily recognizable in the final painting, others are transformed almost beyond recognition, and yet others Picasso tossed out altogether. What is more, Picasso's lover, Dora Maar, photographed various stages of progress in the painting's creation.[2]

These sketches and photographs have been a boon to students of creativity. One of them is psychologist Dean Simonton, who has used the sketches to study the nature of human creativity, specifically the idea that human creativity is a microcosm of Darwinian evolution, played out inside the human mind.[3]

This idea is not new. It is, in fact, much older than Picasso's painting and arguably even older than Darwin's *Origin of Species*. In 1855, four years before Darwin published *Origin*, the Scottish psychologist and philosopher Alexander Bain espoused it when he pointed out that trial-and-error is important for creativity.[4] Twenty-five years after Bain, the philosopher and psychologist William James described the creative process as "a seething caldron of ideas, where everything is fizzling and bobbing about in a state of bewildering activity, where partnerships can be joined or loosened in an instant, treadmill routine is unknown, and the unexpected seems the only law." To him, "the genius of discovery depends altogether on the number of these random notions and guesses which visit the investigator's mind."[5]

But the notion that creativity resembles Darwinian evolution only gained serious traction in 1960, when an article by psychologist Donald Campbell introduced the term *blind variation and selective retention*—often abbreviated as BVSR—which is still widely used to describe it.[6] Its essence is this: just as mutations blindly create genetic variation, we humans blindly create variants of images—or texts, concepts, and ideas. And just as natural selection retains some organisms and discards others, we retain those ideas that are pleasing, useful, or simply apt—like those Picasso selected to convey the horrors of war. One important consequence of Darwinian creativity is that it faces the same towering obstacle as Darwinian evolution: successful creation can require much more than what shortsighted selection can provide.

It is easy to accept that selection matters for Darwinian creativity, because we obviously select some ideas and reject others. In contrast, the blind variation part is harder to swallow. We do not know exactly how new thoughts, concepts, and ideas—useful or not—originate, a bit like Darwin, who did not know how variation originates. The comparison is useful, but it is also misleading. Unlike Darwin, who knew nothing about DNA, we do already know the ultimate substrate of new thoughts. It is the firing of neurons in our brain. The firing rate of neurons fluctuates randomly and spontaneously because neurons sometimes release neurotransmitters at random, exciting other nearby neurons and causing them to fire. Ultimately, these random firings are caused by the same thermal vibrations seen in molecules and atoms—heat—that are responsible for the folding of proteins, the catalysis of chemical reactions, and the self-assembly of crystals.

Here is how Stanislas Dehaene, one of the world's leading neuroscientists, describes this process:

There is nothing magical behind the notion of spontaneous
activity. Excitability is a natural, physical property of nerve
cells.... Fluctuations [in neural activity] are due in large part
to the random release of vesicles of neurotransmitters....
This randomness arises from thermal noise.... What starts
out as local noise ends up as the structured avalanche of spon-
taneous activity that corresponds to our covert thoughts and
goals. It is humbling to think that the stream of consciousness,
the words and images that constantly pop up in our mind and
make up the texture of our mental life finds its ultimate ori-
gin in random spikes sculpted by the trillions of synapses laid
down during our lifelong maturation and education.[7]

So we already know the root and random causes of new
thoughts. What we do not know yet is how exactly the firing of
one (or a trillion) neurons translates into a new thought. Perhaps
it would console neuroscientists that those biologists who study
DNA mutations still struggle with a similar problem. Whenever
a single DNA mutation brings forth a mouse with a new coat
color, a fly with a crippled wing, or a plant with larger leaves,
it can take geneticists years to find out *how* the DNA change
causes the new phenotype. Any one gene cooperates with hundreds or thousands of others, and disentangling their role in the
phenotype remains challenging. Disentangling how neurons
produce conscious thoughts will be no less challenging.

Another reason why people resist accepting creativity as
Darwinian originates from a common misunderstanding: blind
or random variation is often taken to mean creation from
scratch. But that is not its true meaning, not even in biological
evolution, where DNA mutations modify already existing DNA.

Mutate the DNA of a fish, and you get another kind of fish, not a bird, reptile, or dinosaur.[8] (Darwin did not know about DNA, but he recognized the essence of this principle when he coined the phrase "descent with modification.")

As in evolution, so too inside our heads—blind variation does not mean creation from scratch there either. A creative mind does not create arbitrary images that are unrelated to what it is trying to express. To prepare *Guernica*, Picasso's mind did not turn to a playing child, a blooming flower, or a rising sun, which might not have worked to visualize the unspeakable. Nor were Picasso's sketches completely unrelated to previous works of art. The horse, for example, occurs not only in *Guernica*, but is a frequent theme in Picasso's paintings, and the woman with the dead child resembles one of Francisco de Goya's etchings in the *Disasters of War*. In other words, the capacity to create major paintings, novels, or theories requires training and experience so as to create the right kinds of variation. Picasso's mind contained a huge stockpile of imagery before he created *Guernica*—it had to, or a painting as powerful as *Guernica* could not have emerged. Likewise, physicist Paul Dirac had to be steeped in physics and mathematics, or he could not have predicted the existence of antimatter. Dostoyevsky had to be immersed in telling complex stories, or he could not have wrought a masterpiece like *The Brothers Karamazov*. And Beethoven needed to have absorbed innumerable musical phrases before he could compose his symphonies.

In the same vein, because blind variation builds on past experience, Paul Dirac could not have come up with Beethoven's ninth symphony, nor would Dostoyevsky have discovered vaccination. Louis Pasteur's famous saying that "chance favors the prepared mind" applies to much more than scientific discoveries. Preparing a mind requires a life's worth of learning and experiences. These experiences lay down a pattern of neural wiring that guides which new thoughts, images, or melodies can emerge spontaneously—just like a genome's existing DNA restricts what new mutations can create.

So neither creators nor their ideas are blind to the past, but they, like evolving organisms, are blind to something else: the future. Just as nature cannot foresee how a mutant organism will fare, Picasso could not foresee how a particular sketch would fit into the whole painting.[9] Had Picasso been clairvoyant, he could have created *Guernica* in one fell swoop and saved himself all those sketches. No need to paint the warrior's upraised fist, which appears in an early stage of the painting, because it will get modified and later disappear again. And no need to produce sketch number nineteen, the head of a man with bull's horns, nor sketch number twenty-two, a bull with a human head, because neither of them will make it into the final painting.[10]

Other creators are no more farsighted. When University of New Mexico psychologist Vera Steiner-Johns examined the lives of more than one hundred eminent creators—including painter Diego Rivera, chemist Marie Curie, and composer Aaron Copland—she found the telltale signs of such blindness: their minds spouted a profusion of novel ideas before they selected those worth keeping.[11]

Many creators are aware that their ideas are hit or miss. The French poet and essayist Paul Valéry said that "it takes two to invent anything. The one makes up combinations; the other chooses." The English poet and playwright John Dryden described a nascent play more floridly as "a confused mass of thoughts, tumbling over one another in the dark; when the fancy was yet in its first work, moving the sleeping images of things toward the light, there to be distinguished, and then either chosen or rejected by judgment." The physicist Michael Faraday said that his ideas had "been crushed in silence and secrecy by his own severe criticism" and that "in the most successful instances not a tenth of the suggestions, the hopes, the wishes, the preliminary conclusions [were] realized" in his final work. The chemist Linus Pauling said it more succinctly when he argued that a successful scientist must "have lots of ideas and throw away the bad ones."[12] Jacob Rabinow, an inventor of devices as different as optical scanners and pick-proof locks, as well as a member of the National Inventors Hall of Fame, said that "you must have the ability to get rid of the trash which you think of. You cannot think only of good ideas....And if you're good, you must be able to throw out the junk without even saying it. In other words, you get many ideas appearing and you discard them." And John Backus, a computer scientist who helped create the widely used scientific programming language FORTRAN, said that to work successfully, "you need the willingness to fail all the time. You have to generate many ideas and then you have to work very hard only to discover that they don't work. And you keep doing that over and over until you find one that does work."[13]

Such testimonials are only a sliver of the evidence for the Darwinian nature of creativity. Other evidence comes from the serendipitous nature of many scientific discoveries. DuPont chemist Roy Plunckett stumbled upon Teflon when he tried to create a new refrigerant gas. The British inventor Thomas Newcomen discovered the atmospheric steam engine when a broken seam in an engine's outer envelope accidentally injected cold water into the engine's steam cylinder. And Louis Pasteur discovered an important principle of vaccination when he found that a spoiled culture of chicken cholera can immunize chicken.[14] (Such creative accidents also help us see that we tend to overrate human brilliance.)

Spectacular successes like these make history, but at the price of the many failures on which they are built. We can get a glimpse of these failures and their numbers wherever written publications immortalize a creator's work. Such publications—and how often others cite them—provide further evidence for Darwinian creativity. Each citation of a publication pays an intellectual debt to the work. The greater this debt, the greater is the number of citations, and the greater is the work's influence. For this reason, citation counts can quantify not just the influence of a single publication, but that of an entire body of work—and of a creative person. That's also why citation counts are a popular currency of influence when academic or government committees award grants, prizes, and jobs.[15]

At one extreme of the influence scale are publications cited by thousands. They are the best predictors of scientific distinctions such as Nobel Prizes.[16] At the other extreme is work that never gets cited. It is like the proverbial tree that falls in the forest while nobody is looking. And it is astonishing how many trees fall unnoticed, how much effort of creative people goes to waste. Year after year, thousands of newly appearing publications are completely ignored by others, another case in point for creation by trial and—mostly—error. The number of misses is especially egregious in the humanities, where more than 80 percent of papers are not cited even once five years after they have been published.[17] Numbers like this might bring shoddy research to

mind, but even eminent creators, those whose work revolution-
ized their field, also produce inconsequential work. A lot of it.

For example, let's look at the publications of ten influential
psychologists, including luminaries like B.F. Skinner, a father
of behaviorism who showed how to manipulate the behavior of
animals and humans, or Wolfgang Koehler, who taught us that
chimpanzees can solve problems creatively. Of all the publications that these psychologists wrote, 44 percent were not cited
even once within five years after publication.[18] That's almost
half the work of the best scientists in their field, completely ignored. And eminent artists do not fare better—most of their
work seems not to be created for eternity either. For example,
only 35 percent of the work of ten famous composers, including
Mozart, Bach, and Beethoven, is still performed or recorded.[19]

Not only do the greats produce widely ignored ideas, but
they can also make blunders so stunning that they are still remembered a century later. Lord Kelvin, for example, a towering
figure in nineteenth-century physics whose name is immortal-
ized in the scientific unit of temperature, underestimated the age
of the earth by a hundred-fold. His estimate disturbed Charles
Darwin greatly because it meant that evolution would not have
had enough time to create life's diversity. (The estimate was
proven wrong by Ernest Rutherford in the early twentieth century.) No less a genius than Isaac Newton maintained that an
achromatic lens—one that focuses light of different colors on
the same plane—cannot be built. A few centuries later, such
lenses are commonplace in microscopes. And Albert Einstein,
who disavowed quantum theory because he felt that "God does
not throw dice," stubbornly developed a unified theory of physics
that was doomed for that very reason.[20]

If outstanding creativity is hit or miss, yet another historical pattern should follow. Dean Simonton calls it the "constant
probability of failure": the more dirt you pan, the more worthless
gravel you will find.[21] Poet W.H. Auden expressed it this way:
"Chances are that, in the course of his lifetime, the major poet
will write more bad poems than the minor"—simply because major poets write many poems. The flip side is a constant probability of success: the more dirt you pan, the greater your odds
are of hitting gold—the more creative works you produce, the
greater should be their success. This is indeed the case, as one
can prove for scientists by studying citation patterns. To begin
with, a scientist's total number of citations rises with the total
number of publications. That much is perhaps obvious. Less obvious is that the total number of publications also predicts the
number of citations that the scientist's top three publications receive. What is more, the most prominent US scientists—Nobel
laureates—publish on average two times as many papers as their
less prominent peers. Also remarkable is that this pattern does
not just exist for the recent past. It holds all the way back to the
nineteenth century, where the total career output of a scientist
predicts his name recognition to this day.[22] Exceptions do exist,
like the Austrian monk Gregor Mendel, who published little and
was ignored for half a century, but whose experiments on pea
plants eventually triggered the genetics revolution of the twentieth century. But the small number of such exceptions confirm
the rule.

Citation records provide further evidence for the blindness
of human creativity by proving that creativity does not depend
on a person's age. Dirac put it most brutally when he called a

physicist "better dead than living still, when once he's past his thirtieth year."[23] Other fields have different myths about when creativity runs out, but it turns out they are just that: myths. The fraction of nuggets found in mental gravel does not change much over a lifetime. That's what Simonton and others found when surveying areas as different as history, geology, physics, and mathematics. For example, in a survey of more than four hundred mathematicians, work by younger and older mathematicians receives similar recognition.[24] Younger creators do not always find more nuggets than older ones, nor do older creators benefit from greater experience.

A recent analysis of more than two hundred thousand physics publications—and more than half a million articles in fields like biology and economics—also disproves Dirac's callous statement: physicists and other scientists produce important work at any age, with no discernible trend. And what's most important, a creator's best work tends to emerge whenever she produces the most work. That's the constant probability of success all over again.[25]

Lack of foresight. Serendipitous discovery. Many misses. Flops of great minds. Constant probability of success. All these patterns of human creativity underscore the same point. New variants of ideas, images, and concepts are created blindly, in the same sense as new variants of DNA are. Like biological evolution, we are blind to the future success of our creations.

Although selection's role in human creativity is more obvious than that of blind variation, it also leaves room for misunderstanding. Clearly, unless our minds select a useful idea, image, or concept for further elaboration, improvement, or publication, that idea will eventually disappear. Such mental selection is just as essential as natural selection is in biological evolution. However, this does not mean that it is the only force driving creativity, or even just the most important one, for the same reason that natural selection's uphill drive is insufficient for biological evolution: it cannot conquer the complex landscapes of creation, because it is unable to accept inferior solutions that lie on the path to better ones.

Because we know little about how our brains encode ideas, it may be a long time before we can map the landscapes our minds explore. But we are beginning to understand that our brains organize much information about the outside world—even abstract concepts—in some spatial form. For the most immediate information provided by our senses, this has been known for more than a century.[26] Take color, where our minds perceive three major dimensions—hue, saturation, and brightness—such that an object's color occupies a location in a space of colors.[27] One advantage of encoding such information spatially is that objects in a space have a distance, which helps our minds judge instantly whether two colors are similar, like bright orange and bright yellow, or very different, like bright orange and dark purple. Another example is the pitch of a sound, which depends on its frequency and can thus be represented in a single dimension. This is how pitches are mapped onto the linear dimension of our inner ear's cochlea and how they are encoded deeper in the auditory cortex of our brain.[28]

With this knowledge, it takes no great leap to accept that our brains encode other, more complex or abstract concepts, like the sounds of spoken letters, the identity of animals (cat, dog, cow, etc.), or the properties of fruits (color, texture, taste, etc.), in a similar spatial manner. A forceful advocate of this view is Peter Gärdenfors, a Swedish cognitive scientist, who calls such encodings *conceptual spaces* and argues that we need to understand that thought has a *geometry*.[29] He shows that such spaces can help explain how we compare concepts, how we learn new concepts, and how we create new combinations of concepts.

Recent experiments prove that Gärdenfors is on to something. In these experiments, participants were shown a cartoon image of a bird on a computer screen and were taught how to use a computer program that can manipulate two aspects of this bird's body shape: the length of the neck and the length of its legs. The researchers running the experiment trained the participants to use this program to create various bird shapes and to morph these shapes into one another. The researchers wanted to find out how these shapes actually become encoded in the participants' brains, knowing that they *could* be encoded spatially; for example, in a two-dimensional conceptual space of varying neck and leg lengths. The trained participants did not know this, but their brains apparently did. When they watched birds morphing into one another, the same brain regions became activated that people use when navigating physical space, in a pattern that is highly characteristic of such navigation. In other words, not only do our brains encode these bird shapes spatially, but in doing so they piggy-back on the same neural circuitry we use to navigate the world.

The shape of a bird is a pitifully simple concept when compared to the sophisticated methods our minds use to solve complex problems. It may be many years before we know how such ideas are organized in our brain, how a brain explores the mental spaces in which they exist, and how many dimensions these spaces have—I suspect many. Fortunately, these are all small details compared to the most fundamental principle from previous chapters: difficult problems share the fundamental property that their solutions form a rugged landscape. Regardless of how our brains encode these solutions, and whatever space we must explore to find the best ones, we need to overcome this ruggedness. In other words, we can be sure that creative solutions to hard problems anywhere—including those our minds solve—will require finding the high peaks of a rugged landscape in some mental space.

And that's where the steady improvement and hill climbing of selection gets into familiar trouble: it does not allow things to get worse before they get better, and it seals off the valleys that need to be crossed. To cross these valleys, nature reins in the power of selection. It turns out that our minds can do something just like that. Picasso's *Guernica* helps make that point.

In a study that retraced Picasso's journey towards *Guernica*, Dean Simonton presented all of Picasso's forty-five sketches to four independent judges after shuffling them to erase any information about the order in which they had been created. The judges' task was to order these sketches such that the first sketch would be the least similar and the last sketch the most similar to the final painting. Simonton then compared the judges' ordering of the sketches with the actual order in which Picasso had created them. If Picasso's sketches had improved steadily toward the final painting, then the judges would order the sketches in the same order in which Picasso had created them.

But they didn't. Ordered by resemblance to the final painting, the sketches were scrambled with respect to creation date.
Picasso's mind did not climb a single hill toward his final painting. His path would lead uphill only to go down again, moving
in multiple zig-zags toward the final result. Some of his sketches
are similar to a figure in the final painting, only to be followed by
others that bear little resemblance. Some sequences of sketches
appear to steadily improve toward the final painting, only to
drop into a valley of little resemblance. What is more, the path
zig-zagged also in sequences of sketches experimenting with the
same motif—the woman with the dead child, the fallen warrior,
or the screaming horse.[30] And some motifs didn't make it into
the final painting at all, such as the outstretched fist of the warrior, which became transformed in later versions and eventually
disappeared.[31]

Simonton's systematic study quantifies what creative people have known all along: the path toward a creative product
is neither straight nor all uphill. Perhaps Rainer Maria Rilke
referred to this path when he described in dark imagery that a
poet must "have been among the shades. . . . You have to sit down
and eat with the dead."[32] His words echo mythological journeys
into the underworld, like those of Orpheus, Virgil, and Dante—
themselves powerful metaphors for the trials of creation. Writer
Margaret Atwood puts it like this: "Poets travel the dark roads.
Inspiration is a hole that leads downward."

The nineteenth-century mathematician and polymath Henri
Poincaré—a father of chaos theory—connected his creative
journey to another familiar one. It's the one nature uses to create molecules like bucky-balls, which find the deepest valleys in

their energy landscapes but not before traversing many shallow valleys and saddles with unstable molecules. For instance, Poincaré describes one sleepless evening when "ideas rose in crowds; I felt them collide until pairs interlocked, so to speak, making a stable combination."[33] And according to the French philosopher Paul Souriau, ideas arranged in a creator's mind by chance are "shaken up and agitated..., form numerous unstable aggregates which destroy themselves and finish up by stopping on the most simple and solid combination."[34] Remarkably, both statements predate the concept of an energy landscape by decades.[35]

And recall the testimony of nineteenth-century physicist Hermann von Helmholtz, who compared his progress in solving a problem to that of a mountain climber "often compelled to retrace his steps because his progress stopped" and who "hits upon traces of a fresh path, which again leads him a little further."[36]

These introspections bring us back to an already familiar question: how *do* creative minds overcome those valleys to get to the next higher peak?

Chapter 8

Not All Those Who Wander Are Lost

B ecause thinking minds are different from evolving organisms and self-assembling molecules, we cannot expect them to use the same means—mechanisms like genetic drift and thermal vibrations—to overcome deep valleys in the landscapes they explore. But they must have some way to achieve the same purpose. As it turns out, they have more than just one—many more.

Let's begin with play.

I don't mean the rule-based play of a board game or the competitive play of a soccer match, but rather the kind of freewheeling, unstructured play that children perform with a pile of LEGO blocks or with toy shovels and buckets in a sandbox. I mean playful behavior without immediate goals and benefits, without even the possibility of failure.

Play is so important that nature invented it long before it invented us. Almost all young mammals play, as do birds like parrots and crows.[1] Play has been reported in reptiles, fish, and even spiders, where sexually immature animals use it to practice copulation. But the world champion of animal play may be the bottlenose dolphin, with thirty-seven different reported types of play.[2] Captive dolphins will play untiringly with balls and other toys, and wild dolphins play with objects like feathers, sponges, and "smoke rings" of air bubbles that they extrude from their blowholes.

Such widespread play must be more than just a frivolous whim of nature. The reason: it costs. Young animals can spend up to 20 percent of their daily energy budget goofing around rather than, say, chasing dinner. And their play can cause serious problems. Playing cheetah cubs frequently scare off prey by chasing each other or by clambering over their stalking mother.[3] Playing elephants get stuck in mud. Playing bighorn sheep get impaled on cactus spines. Some playful animals even get themselves killed.[4] In a 1991 study, Cambridge researcher Robert Harcourt observed a colony of South American fur seals. Within a single season, 102 of the colony's pups were attacked by sea lions, and twenty-six of them were killed. More than 80 percent of the killed pups were attacked while playing.[5]

With costs this high, the benefits can't be far behind. And indeed, where the benefits of play have been measured, they can make the difference between life and death. The more feral horses from New Zealand play, for example, the better they survive their first year.[6] Likewise, Alaskan brown bear cubs that played more during their first summer not only survived the first winter better, but also had a better chance to survive subsequent winters.[7]

Some purposes of such play have nothing to do with mental problem solving. When horses play, they strengthen their muscles, and that very strength can help them survive. When lion cubs play-fight, they prepare for the real fights that will help them dominate the group. When dolphins play with air bubbles, they are honing their skills at confusing and catching prey. And when male spiders play at sex, they practice how to copulate fast enough to get away from a female before other males attack them.[8]

But at least in mammals, play goes beyond mere practice of a stereotypical behavior, like that of a pianist rehearsing the same passage over and over again. When mammals stalk, hunt, and escape, they find themselves in ever-new situations and environments. Marc Bekoff, a researcher at the University of Colorado and a lifelong student of animal behavior, argues that play broadens an animal's behavioral repertoire, giving them the flexibility to adapt to changing circumstances. In other words, animal play creates diverse behaviors, regardless of whether that diversity is immediately useful. It prepares the player for the unexpected in an unpredictable world.[9]

That very flexibility can also help the smartest animals solve difficult problems. A 1978 experiment demonstrated its value for young rats. In this experiment, some rats were separated from their peers for twenty days by a mesh in their cage, which prevented them from playing. After the period of isolation, the researchers taught all the rats to get a food reward by pulling a rubber ball out of the way. They then changed the task to a new one where the ball had to be pushed instead of pulled. Compared to their freely playing peers, the play-deprived rats took much longer to try new ways of getting at the food and solving this problem.[10]

University of Cambridge ethologist Patrick Bateson linked observations like this more directly to the landscapes of creation when he argued that play can "fulfill a probing role that enables the individual to escape from false endpoints, or local optima" and that "when stuck on a metaphorical lower peak, it can be beneficial to have active mechanisms for getting off it and onto a higher one."[11] In this view, play is to creativity what genetic drift is to evolution and what heat is to self-assembling molecules.

If that is the case, it is hardly surprising that creative people often describe their work as playful. Alexander Fleming, who would discover penicillin, was reproved by his boss for his playful attitude. He said, "I play with microbes. . . . It is very pleasant to break the rules and to find something that nobody had thought of."[12] Andre Geim, 2010 Nobel laureate in physics, declared that "a playful attitude has always been the hallmark of my research. . . . Unless you happen to be in the right place and the right time, or you have facilities no one else has, the only way is to be more adventurous."[13] When James Watson and Francis Crick discovered the double helix, they had help in the form of colored balls they could stick together—LEGO-like—to build a model. In Watson's words, all they had to do was "begin to play." And C.G. Jung, one of the fathers of psychoanalysis, said it best: "The debt we owe to the play of imagination is incalculable."[14]

One hallmark of play is that it suspends judgment so that we are no longer focused on selecting good ideas and discarding bad ones. That's what allows us to descend into the valleys of imperfection to later climb the peaks of perfection. But play is only one means to get there.

Less deliberate but just as powerful are the dreams that we experience in our sleep. It is no coincidence that the psychologist Jean Piaget, whose trailblazing research helped us understand how children develop, likened dreaming to play.[15] It is in dreams that our minds are at their freest to combine the most bizarre fragments of thoughts and images into novel characters and plotlines. Paul McCartney famously first heard his song "Yesterday" in a dream and did not believe that it was an original song, asking people in the music business for weeks afterward whether they knew it. They didn't. "Yesterday" would become one of the twentieth century's most successful songs, with seven million performances and more than two thousand cover versions. Another dream whispered to the German physiologist Otto Loewi the idea for a crucial experiment, which proved that nerves communicate through chemicals that we now call neurotransmitters. It would win him a Nobel Prize.

Even in the state of half-sleep—psychologists call it hypnagogia—our minds are sufficiently loose to descend from those lowly hills. In this state, August Kekulé saw the chemical structure of benzene, Mary Shelley found the idea for her iconic novel *Frankenstein*, and Dmitri Mendeleev discovered the periodic table of the chemical elements.[16]

Similar to playing and dreaming is the wandering of our minds. Ninety-six percent of adult Americans report that it happens to them daily—the other 4 percent may be too absentminded to notice. To quantify how often any one mind wanders during a task is simple: ask. Interrupt people who work on the task and ask what's on their mind. Or let mobile phones do the work for you. Program them to send study participants a text asking what they are thinking about at random times of the day.[17] When psychologists do that, they find that mind-wandering is staggeringly frequent. The typical mind is absent between a third and half the time.[18]

Mind-wandering is often considered a harmless quirk, as in the cliché of the scatterbrained professor. But it has real consequences. Let's begin with the bad ones. Absentminded people perform less well on tests that require focused attention, such as reading comprehension tests. More worrisome, they also perform more poorly on tests that you better not flunk if you have any career aspirations. Among them is the Scholastic Aptitude Test (SAT) that many colleges require for admission.[19]

But mind-wandering also has an upside—at least for well-trained minds. Indeed, many anecdotes of creators like Einstein, Newton, and Poincaré report that these scientists solved important problems while not actually working on anything. The common wisdom that the best ideas arrive in the shower is exemplified by Archimedes's discovery of how to measure an object's volume. (Ok, he was in a bathtub.) But while Archimedes's discovery was triggered by the rising water as he entered the tub, other breakthroughs surface apropos of nothing. Take this well-known quote from the eminent mathematician Henri Poincaré describing a period in his life when he had worked without success on a mathematical problem:

> *Disgusted with my failure, I went to spend a few days at the seaside, and thought of something else. One morning, walking on the bluff, the idea came to me, with . . . brevity, suddenness, and immediate certainty, that the arithmetic transformations of indeterminate ternary quadratic forms were identical with those of non-Euclidean geometry.*[20]

The apparently idle period before such insights arrive has a name: incubation. If hard and seemingly futile work on a difficult problem is followed up with a less demanding activity that does not require complete focus—walking, showering, cooking—a mind is free to wander. And when that mind incubates the problem, it can stumble upon a solution.

Incubation is as unconscious as it is real, and it enhances creativity. In one experiment making that point, 135 college students took a psychological test for creativity that required them to find unusual uses for everyday objects, like bricks or pencils. A few minutes into the test, the psychologists running the experiment interrupted some students and gave them an unrelated task. The new task did not take much effort—the students were shown a series of digits and had to tell which of them were even or odd—but it distracted the students from the test. After that interruption, the students continued with the creativity test, and they found more-creative answers than a second group of students who had not been given the distracting task.

Students in a third group got a break like the first, but they were given a harder task that required more focus. And, lo and behold, their answers were less creative than those of the first group. The conclusion: undemanding tasks—easy enough to require little attention, but hard enough to prevent conscious work on a problem—can free a mind to wander and solve a problem creatively.[21]

If mind-wandering impacts creativity, then its opposite, the control of attention practiced in mindfulness meditation, should have the opposite effects, both good and bad. And indeed it does. A 2012 study showed, for example, that mindfulness meditation, by reducing mind-wandering, can improve scores on standardized academic tests.[22] In contrast, less mindful individuals perform better on creativity tests like that just mentioned.[23]

The message is clear: just as biological evolution can require a balance between natural selection, which pushes uphill, and genetic drift, which does not, so too does creativity require a balance between the selection of useful ideas—where a focused mind comes in handy—and the suspension of that selection to play, dream, or allow the mind to wander.[24]

The importance of reaching a state of mind—sometimes—where selection is suspended is best illustrated by the many ways in which humans try to attain this state.

Some means to this end are as simple as creating a playful environment.[25] When companies that value their "creatives" provide the wacky workspaces made famous by Google—complete with slides, firemen's poles, hammocks, and foosball tables—they aim to create the kinds of environments we remember from the playgrounds of our childhoods.

A great idea, to be sure, but, sadly, toys alone will not transform office drones into brainstormers. One problem is that we are highly attuned to the judgment of others. This judgment, like the judgment we impose on our own thoughts, is a form of selection that punishes failure. Fear of it creeps into our psyche at some point on our path to adulthood, and Stanford researcher Robert McKim showed how pervasive it can be. In a simple experiment, he had students in a classroom draw a neighbor's portrait in thirty seconds and show the portrait to the neighbor. Most of the students felt embarrassed about their drawings and apologized to their "victims."[26] No longer as guileless as children, who will proudly exhibit their latest opus to anybody, they had come to expect their neighbors' criticism.

Resurrecting the inner child in an adult's mind can require special measures. Among them are the ground rules of brainstorming sessions, including "don't criticize others" and "don't compare ideas." To push against the habit of judging, some companies even award prizes for the most outlandish ideas in their brainstorming sessions.[27]

But these vehicles to cross the valleys of inferior ideas have speed limits. To go faster and farther, some creators use more forceful means—like drugs. The vivid pipe dreams of opium smokers have triggered major creative works, including Samuel Taylor Coleridge's celebrated poem "Kubla Khan." Nobel Prize–winning biochemist Kary Mullis credited LSD for his invention of a technique to copy DNA molecules. Steve Jobs called his LSD experience "one of the two or three most important things I have done in my life" and was fond of asking aspiring creative professionals whether they had ever dropped acid. And his generation was not the first to appreciate the power of psychedelic drugs. Works of art created thousands of years before the Macintosh computer have been linked to the use of psychedelic drugs in the prehistoric cultures of Europe, Africa, and South America.[28]

Psychologists began to study the effects of psychedelic drugs on thinking in the 1960s. One study focused on twenty-seven workers in creative professions like engineering, design, physics, architecture, and art. Each took a creativity test before and another one after a dose of mescaline and, while drugged, worked on a problem from his or her area of expertise, such as designing a commercial building, an improved magnetic tape recorder, or a piece of furniture. The participants not only fared better on the creativity test while under the influence, but they also felt that mescaline improved their creative problem solving, both during the experiment and for two weeks thereafter. One subject described exactly the kind of relaxed condition that enables creativity: "no fear, no worry, no sense of reputation and competition, no envy." Another said, "I began to draw.... My senses could not keep up with my images.... My hand was not fast enough.... I worked at a pace I would not have thought I was capable of." And yet another testified that his mind "seemed much freer to roam around the problems, and it was these periods of roaming around which produced solutions."[29]

Unfortunately, this and other such studies do not meet the more rigorous standards of today's psychology. For example, they do not compare the drug-takers' creativity with a group of control subjects that did not take the drug. So, the jury is still out.

Fortunately, other drugs might boost creativity, even if LSD doesn't. The Romans knew this when they said, "There is no poetry among water drinkers."[30] Science is still catching up with such folk wisdom, although one study made a similar point when it found that twenty mildly intoxicated social drinkers did better on a creativity test than did twenty sober participants.[31]

But whether any one drug is best at enhancing creativity matters less than this: there is more than one state of mind—and more than one way to find it—where judgment and the selection of ideas are suspended. That is a good thing, because the ability to descend into the depths of half-baked sketches, inferior drafts, or imperfect harmonies is essential to reach the heights of a great work. Play, dreams, and the still mysterious ways of incubation join the genetic drift of adaptive landscapes and the thermal vibrations of energy landscapes as means of overcoming obstacles in these landscapes.

And they hint at another talent of creative people, whose biological counterpart has helped biological evolution succeed: not only are their minds well traveled, but they can also move rapidly—yes, even teleport instantly—through a mental landscape.

Much of creation is a journey through an abstract, high-dimensional realm, but creative works can also build on journeys in a more familiar world. This is especially true for the journeys of outstanding individuals, the trailblazers of humankind. I mean people like the French artist Paul Gauguin.

Born in Paris, Gauguin spent most of his early childhood in Peru, where his mother's interest in pre-Columbian pottery became his earliest artistic influence. But his life would take a long detour before returning to art. His family moved to France in 1855, when Gauguin was seven years old, and he entered a naval preparatory school a few years later. He entered the merchant marine, traveled the oceans for three years, and later served for two years in the French navy. Upon returning to Paris at age twenty-three, he worked as a stockbroker for eleven years, until that career was derailed by an 1882 stock market crash. He then moved to Denmark with his Danish wife and tried himself as a tarpaulin salesman. Unfortunately, the Danish did not take a liking to French tarpaulins, so his wife had to support the family by giving French lessons.[32] During his time as a stockbroker, Gauguin had started to paint, and with his commercial career in shambles, he decided to paint full-time. To make a go of it, he left his family in Denmark and moved back to Paris. The going was rough. His work achieved neither widespread critical acclaim nor financial success, and Gauguin had to take on menial jobs. Disappointed with the French art scene, he eventually left France, first for Panama and Martinique, and then for Tahiti and the Marquesas Islands in the South Pacific. Living in a bamboo hut for a while, he produced his best-known work on these islands. These are highly valued—and highly priced—paintings like *When Will You Marry* or *Ave Maria*, which depict Polynesian natives in bold colors and surrounded by lush tropical sceneries.

The life journey of Renaissance painter Raphael did not range as widely, but his travels—from the studio of his father, where he first learned to paint, to the studio of Pietro Perugino, whose style and method of glazing he assimilated, to Florence, where he absorbed aspects of da Vinci's *sfumato* and pyramidal constructions—helped him create a new style he called *unione*, which is embodied in his masterworks, like *The Transfiguration of Christ*.

The creative journeys of many other creators are lost in time, even though they helped create entirely new artistic styles. Among them is a style of painting popular in Renaissance Venice, which amalgamates Byzantine Art—stone mosaics, flat-panel paintings—with Western elements like the three-dimensional renderings of linear perspective.[33] In Latin America, the facades of famous Baroque churches like that of San Lorenzo in Bolivia's Potosí merge Christian symbols like angels with those of the Incan religion, like sun and moon, in a style that art historians call Andean Hybrid Baroque.[34] And in Western architecture, the pointed arch helped transform squat Romanesque churches into light-flooded Gothic cathedrals that stretch into the sky.[35] This arch had been used for centuries in the Islamic architecture of the Near East, so chances are that it arrived with some itinerant engineers or masons.

These and other new styles of art emerged when creators both known and unknown traveled the world. But an itinerant life as such isn't the point. The point is that such a life can enable an inner journey through different realms of knowledge. A great example of such an inner journey is the convoluted path of Ilya Prigogine, who won the 1977 Nobel Prize in chemistry but did not start out with scientific aspirations. His first love was the humanities, most notably philosophy. Perhaps there are parents who rejoice when their son wants to become a philosopher, but Prigogine's were not among them. They insisted that he pursue a more reputable line of work. So he studied law. Unfortunately, law was not his true calling, but during his studies he became fascinated by the criminal mind and its psychology. To grasp the mind's hidden springs of action, Prigogine decided he needed to understand brain chemistry. That endeavor turned out to be too ambitious for the times he lived in, so he turned to simpler chemical systems that display self-organization—the process by which systems as different as cyclones and viruses spontaneously self-assemble. And that's where he made his mark in science, discovering laws of nature that make self-organization not only possible but inevitable. Eventually, his scientific work led him all the way back to philosophy and to the question of whether the world is deterministic and whether choice, responsibility, and freedom are illusory concepts. He came out against determinism. His arguments were based on his research in chemistry, where he had shown that the future of some chemical systems cannot be predicted with certainty. Prigogine's life's work, an example of outstanding creativity at the intersection between the humanities and science, was only made possible by his meandering path through different fields of knowledge.[36]

Another well-documented inner journey is that of physicist Rosalyn Yalow, who won the Nobel in the same year as
Prigogine. Her life was pulled by two crosscurrents of intellectual influence.[37] Growing up in the 1930s, when the world celebrated the spectacular successes of quantum mechanics, Yalow
was drawn to physics, and in particular to the radioactivity that
another famous female scientist, Marie Curie, had helped the
world understand decades earlier. Yalow got her first big break
when she was admitted to graduate school in physics at the
University of Illinois, where openings were abundant because
many young men had gone off to fight in World War II. She
was the only woman in a department of more than four hundred. In 1945, right after she received her doctorate, she crossed
over into a new field when she was hired by the radiotherapy
department of the Bronx Veterans Administration Hospital.
She was blissfully ignorant of both medicine and biology, but
proved a quick study who realized how radiation physics could
help medicine. Together with her colleague Solomon A. Ber-
son, she created the first department of "nuclear medicine"—an
entirely new field of science. Her biggest, Nobel Prize–winning
achievement was to invent the radioimmunoassay, a highly sensitive measurement technology that uses radioactive isotopes to
quantify minute amounts of molecules, such as insulin, in a patient's blood. Had she stayed on the straight and narrow of her
traditional physics education, she would never have made this
outsized contribution.

Other examples include the Austrian physician Karl Land-
steiner, whose knowledge of chemistry helped him discover the
main blood groups, which made modern blood transfusions possible. They also include Hermann von Helmholtz, whose father

obliged him to marry medicine instead of his first love, physics. It led von Helmholtz to invent the ophthalmoscope, a device that can illuminate the eye's interior and remains to this day one of the most widespread medical instruments.[38] And August Kekulé of benzene fame abandoned architecture for chemistry, but his fascination with geometry led him to investigate the spatial organization of molecules. His inner journey helped him earn the title "architect of organic chemistry."[39]

Lives such as these help a mind accumulate unusual combinations of experience and expertise. And they permit what philosopher Arthur Koestler's book *The Art of Creation* called cross-fertilization, an amalgamation of knowledge that resembles the kind of recombination we found in biological evolution.[40]

Some of this cross-fertilization takes eminent creators far outside their area of expertise, where their intelligent naïveté can lead to important discoveries. Louis Pasteur had never worked with silkworms before his research rescued the French silk industry from a deadly parasite that decimated its silkworm populations. Henry Bessemer, the inventor of a cheap process to manufacture steel, credited his success to his ignorance of well-established truths. And Luis Alvarez, a nuclear physicist, discovered that a gargantuan asteroid had helped eradicate the dinosaurs—a discovery that was as crucial to paleontology as it was unwelcome to turf-conscious paleontologists.[41]

Koestler's cross-fertilization can be especially powerful when it extends beyond different branches of science to the arts. A well-known example is that of Steve Jobs, whose companies Apple and Pixar melded digital technology and design into an indivisible whole. Jobs amalgamated digital electronics with influences as different as the simple elegance of Bauhaus architecture, the minimalism of Zen Buddhism, and the art of calligraphy to help design revolutionary products like the Macintosh computer. The Mac's elegant proportional fonts, for example, abandoned the clunky fixed-width fonts that computers had used until then and paved the way for the revolution of desktop publishing that everybody takes for granted now—in no small part thanks to Jobs's love of calligraphy.[42] Albert Einstein is also well known for his artistic sensibility. He loved to play the violin and thought that important scientific theories fused the true with the beautiful. He credited music for his discovery of relativity theory when he said that the theory "occurred to me by intuition, and music is the driving force behind this intuition....My new discovery is the result of musical perception."[43] An artistic theme also resounds in the particle accelerators that the physicist and accomplished sculptor Robert R. Wilson conceived: "In designing an accelerator I proceed very much as I do in making a sculpture.... The lines should be graceful, the volumes balanced."[44]

Broad surveys of scientists support the notion of a deep connection between great science and artistic sensibilities. In 2008, Robert Root-Bernstein and colleagues compared the artistic side interests of thousands of scientists and the general public.[45] The most eminent group they studied were the Nobel laureates, followed by scientists just one rung below. These included the Fellows of the Royal Society of the United Kingdom, an organization of more than one thousand British scientists elected for lifetime membership based on their outstanding accomplishments, and members of the National Academy of Sciences of the United States, another body of elite scientists. Root-Bernstein's 2008 study found that twice as many National Academy and Royal Society members exercise an art or a craft as less eminent scientists or the general public. But even they are beaten hands down by the Nobel laureates, where an additional 50 percent have artistic side interests that range from music, sculpture, and painting to fiction writing.

More generally, systematic studies show that many eminent creators have had wide-ranging interests since childhood, which helped them acquire the broad knowledge they needed to cross-fertilize different areas of science or the arts. In contrast, they were often indifferent to the schooling intended to launch the average pupil into the job market.[46] Mark Twain expressed this contrast best when he quipped, "I have never let my schooling interfere with my education."[47]

Unfortunately, creativity requires more than the education afforded by inner journeys like that of Prigogine or by outer journeys like that of Gauguin. Countless people travel the world, explore different professions, or study diverse disciplines, but we do not hear from most of them. They lack a talent that is crucial for creativity.

Here is a clue about the essence of this talent:

A man and his wife, who normally quarrel a lot, spend an unusually harmonious evening at a restaurant. While they enjoy their dinner, a waitress suddenly drops a tray stacked with dishes, which shatter with an earsplitting racket. "Listen, honey!" the husband says. "They are playing our song."

Explaining a joke is a surefire way to destroy it, but it also helps connect humor to a grander theme that we first

encountered in biology. Humor links concepts in a mental landscape that are as remote as the harmony of a love song and the
frightful noise of shattering dishes, a bit like what evolution does
when it recombines very different genes. And when these concepts combine in a good joke, a brief flash of surprise and delight
is created. This is another insight of Koestler, who compared the
creativity of humor, art, and science in the *Act of Creation*. He
identified the energy source common to both the spark of a joke
and the supernova of a major creation when he concluded that
all human creativity is powered by such combinations.[48]

A creative mind must not just absorb wide-ranging concepts,
ideas, or images. It must recombine them, however remote they
may seem. Such recombination can use different vehicles. An
especially powerful one is the analogy. When physicist Max
Planck combined the unrelated concepts of atoms and vibrating
strings into one of the most fertile analogies ever, he became able
to explain that atoms emit energy in discrete, quantized packets.
Louis de Broglie, another father of quantum theory, imagined
that elementary particles like electrons not only vibrate but also
produce harmonics like real strings. His insights paved the way
to modern medical technologies like magnetic resonance imag-
ing. (It may have helped that Planck was a brilliant pianist and
de Broglie an accomplished violinist.)

As in science, so too in engineering. Johannes Gutenberg
combined woodblock printing with the idea of a coin punch to
create movable type, and then combined *that* new idea with the
squeezing action of a wine press. The end result was a revolution in how we preserve and spread knowledge.[49] Surgical staples originated with indigenous peoples on several continents
who closed gaping wounds by letting large ants bite the opposite

edges of these wounds and draw the edges together with their powerful jaws. Velcro was inspired by the clinging of burs to fur and clothing.[50] In the words of economist Brian Arthur, entire technologies "come into being as fresh combinations of what already exists."[51]

By linking seemingly unrelated concepts and phenomena, analogies have helped us discover the most profound laws of nature and create technological revolutions. Their power is similar to that of another element of language, one as central to literature and poetry as analogies are to science. I am thinking of the metaphor. A poetic metaphor like *atoms are ringing bells* expresses the similarity between atoms and ringing bells more succinctly than any science text could.

The word *metaphor* literally means to carry something from one place to another.[52] Metaphors have been essential to creative writing and speaking since the time of Aristotle, who first expounded their power in his *Rhetoric*.[53] Robert Frost declared that he strived to "make metaphor the whole of thinking," and others argue that it already is almost the whole of our thinking.[54] In the words of psychologist Steven Pinker, "Metaphor is so widespread in language that it's hard to find expressions for abstract thoughts that are not metaphorical."[55] We say that a bus stops *at* the station, but also that we meet *at* 6 o'clock, using space as a metaphor for time. We say that a man *goes* to the store, but also that a screen *goes* blank, using movement as a metaphor for change. We do not even recognize such pervasive metaphors for what they are.

Metaphors unearth commonalities between two remote concepts, but the metaphor the "stench of failure" works not just because it connects the unpleasant natures of filth and failure.

Metaphors are more powerful than that, as psychologists Roger Tourangeau and Lance Rips showed in experiments that analyzed the responses of eighty different people to multiple metaphors. Some of the metaphors the scientists had invented themselves, such as "the eagle is a lion among birds." Others they lifted from published poems, such as Robert Frost's *An Old Man's Winter Night*, which contains the line "a light he was to no one but himself." Each study participant was asked to list properties of the different concepts, such as eagle and lion, that occurred in the same metaphor. After that, she had to report which images the whole metaphor conjured in her mind. Tourangeau and Rips found that metaphors do not just lay bare the commonalities between concepts—eagles and lions stalk, they prey on their quarry, and so on. That's because people do not take phrases like "the eagle is a lion among birds" or "a light he was to no one but himself" literally. To do so would be absurd—an eagle is not a lion, nor does an old man shine like a lamp. Instead, the scientists found that metaphors conjure meanings that are new and different from those contained in their parts. For example, Frost's "a light he was to no one but himself" evokes the man's isolation and solitude.[56]

British poet Ruth Padel is referring to this power of mental recombination when she calls metaphors the most extreme movement powering a poem's journey. Spanish poet Federico Garcia Lorca does the same thing when he calls a metaphor an equestrian leap that unites two worlds.[57] Metaphors are the most compact form of recombinant thought.

We can investigate the role of mental leaps in creativity not just by listening to poets or studying metaphors, but also by having people—from high school students to professional artists, scientists, writers, and engineers—take creativity tests. A bewildering variety of such tests exist, but remarkably, all of them measure the same ability in various guises: how far and how fast a mind can travel.

The history of these tests begins in World War II. That's when the US Air Force studied how pilots react during an emergency like a failing plane and found that its smartest pilots did not always react in the best possible way to save their planes and their lives.[58] Instead, the most creative pilots did, and the smartest pilots were often not the most creative.[59] To identify those most creative pilots, the Air Force hired psychologist Joy Paul Guilford to help develop experimental tests for creativity. In the course of this work, Guilford had a simple yet enduring insight. There are two fundamentally different kinds of thinking: divergent thinking and convergent thinking.[60] Divergent thinking is the heart of creativity, providing multiple candidate solutions for a problem, like that of saving a plane, whereas convergent thinking winnows these variants down to a single, hopefully best solution. Indeed, Guilford's work provided further support for the Darwinian nature of mental creativity, where divergent thinking is analogous to processes like mutation and recombination, and convergent thinking is analogous to natural selection. Guilford also realized something else—people's minds differ in how they create new thoughts. Some move in small steps akin to point mutations. Others—the more creative ones—take bigger leaps.

Guilford's ideas helped him develop a variety of tests for divergent thinking, which spawned a whole industry of creativity testing. Among the earliest tests were word association tests, where a person is given a word like *hand* and asked for a list of related words. Where an unimaginative person would respond only with the obvious—*arm*, *foot*, and *finger*—a creative person would come up with more original responses, like *soft*, *friendship*, and *instrument*. A person's responses to each word in such a test would be compared to responses from hundreds or thousands of people. From their responses, we know that *fruit* is close to *apple*, *edible*, and *tree*, but more distant from *medicine*, *garden*, and *wine*. And *butterfly* is close to *insect*, *caterpillar*, and *bird*, but more distant from *blossom*, *sunshine*, and *temporary*.[61]

Such tests also distinguish fluency—the ability to find many related words—from originality—the ability to find unusual words. Not all fluent minds are also original. Some find many associations, but all of them are near each other. Original minds, however, find more distant or remote associations. They are like globetrotters compared to homebodies, who only putter around their backyard.[62]

The notion of distant travel is baked into the very name of another, more sophisticated word-association test. It is the remote-associates test, which starts from a triplet of words like *cottage*, *swiss*, and *cake*. The task is to find a single word that links all three.[63] Most people easily find the solution to this triplet—*cheese*—but more remote triplets are harder to solve. They include *river*, *note*, and *account* (linked by *bank*), as well as *fur*, *rack*, and *tail* (linked by *coat*). The test-taker needs to solve as many triplets as possible within a given amount of time.[64]

Because not all creativity is about words, psychologists also developed tests to quantify nonverbal creativity. One such test is the alternative-uses test, where people are given an object like a brick, a hairpin, or a cardboard box and are asked to come up with as many different uses as they can—a brick could be used as a bookend, a paperweight, a doorstop, a hammer, and so on. In another test, subjects are shown a simple shape, such as a circle, and are asked to draw as many objects with this shape as they can—a sun, a face, a flower, a football, and so on. In yet another test, subjects are asked to imagine the consequences of unusual events, such as if everybody went blind, or if clouds in the sky suddenly had ropes dangling from them, reaching to the ground.[65] The most widely used of these test batteries is the Torrance Test of Creative Thinking, named after its designer, psychologist E. Paul Torrance.[66] (Torrance knew about the importance of suspending judgment, which we heard about earlier, because he emphasized that his tests should be taken in a relaxed atmosphere so that subjects can have fun with the test questions.[67])

These tests are simple, but they work.[68] For example, architecture students at the University of California whose professional work was ranked by their professors as being especially creative also scored highly on a remote-association test.[69] Such expert ratings of creativity may seem a tad subjective for a scientific experiment. However, not only do different raters often agree to a surprising extent, ultimately *all* creative works are evaluated by other humans.[70] Even better than such snapshots of people's creativity are *longitudinal* studies that follow a person's creative output over many years. Two such studies tracked elementary and high school students for up to twenty-two years after they had taken the Torrance test battery.

One of these students is Ted Schwarzrock, who took the Torrance test as an eight-year-old and astonished psychologists with the twenty-five creative improvements he found for the fire truck they handed him. Five decades later, he was a wealthy entrepreneur who had invented medical technologies like respirators and anti-inflammatories.[71] And he was no exception. Students who had done well on the test frequently became more prolific creators, based on criteria such as award-winning artworks, plays, and musical compositions that were publicly performed, as well as patents and inventions.[72]

But creativity tests are not perfect. We know this because many good test-takers do not become prolific creators. And that should also come as no surprise. For one thing, most creativity tests focus on originality but not appropriateness—the wilder an idea, the better. Also, they quantify the talent for divergent thinking to generate ideas, but not that for convergent thinking to isolate solutions. Most of us know the kind of person who has too much of one but too little of the other. Just like biological evolution, human creativity needs a balance between opposing forces.

In addition, typical creativity tests are completed in mere minutes or hours, and without special equipment or skills. That's a good thing—a test asking for a sculpture to be created from a slab of marble would find few takers. But it also creates another weakness because such tests neglect the fact that outstanding creators undergo years of training, and that they may have to endure even longer periods of obscurity. Creativity tests are silent on the grit it takes to stagger through a desert in search of an oasis.[73]

But, warts and all, creativity tests have taught us a lot about creative minds. Their ability to quantify "remoteness" and "closeness"—for example, through frequent word associations—shows that distance can be more than a metaphor for mapping our minds.[74] And they all show that creative minds can traverse great distances. Instantly.

Creativity tests are not alone in making this point. Experiments that help our minds recombine images do as well. In these experiments, psychologist Albert Rothenberg showed pairs of very distant—really, unrelated—images to each member of a group of visual artists or art students. For example, the first image might show several soldiers with rifles crouched near a tank, whereas the second image might show an ornate French four-poster bed. To another group of students, Rothenberg presented a single composite image, with both the soldiers and the bed superimposed. He then asked the artists to create new drawings inspired by these images and had these drawings evaluated by independent experts, which included a professional artist, an art teacher, and an art critic. The result: the composite image evoked more creative drawings.[75]

Rothenberg's composite image does for visual artists what metaphors do for readers of poetry: it launches a leap of the imagination. And the same principle—Rothenberg calls it homospatial thinking—also enhances other forms of creativity. We know this because Rothenberg spent more than a thousand hours interviewing American and British writers and scientists, who confirmed that imagining different kinds of objects in the same space helps their creative work.[76] Rothenberg's work also sheds light on multiple well-known historical anecdotes, such as the one where Kekulé dreamed of atoms morphing into snake-like chains, and when these snakes bit themselves in the tail, he discovered the celebrated ring of benzene's molecular structure.[77] Albert Einstein famously imagined himself traveling along a beam of light when he developed relativity theory, and physicist Donald Glaser envisioned himself in Earth orbit above a tub of liquid hydrogen when he invented a bubble chamber to detect subatomic particles.[78]

Yet other psychological experiments show that creative minds extend their feelers further into the world. And I do not mean that creative people deliberately seek new experiences—that's almost a truism, because you can only recombine what you have experienced. I mean an altogether more subtle, unconscious sensibility.

In one such experiment, 204 college students had to solve anagrams—puzzles where letters in a word like *senator* can be rearranged to yield other words, like *treason*. Before this task, the participants were shown a list of words, and at the same time, another list of words was read to them aloud. They were asked to memorize the words they read, but to ignore the words they were hearing. What they didn't know is that both kinds of words included solutions to some of the anagrams they would have to solve.

When they did solve the anagrams, highly creative students differed from the rest: they did better on those anagrams whose solutions they had heard but were asked to ignore. In other words, their minds were less scrupulous at blending out supposedly irrelevant but actually crucial information.[79] This pattern will strike a chord in anybody familiar with the history of scientific discovery, which is full of people who pay attention to information ignored by others. In the words of chemist Albert Szent-Györgyi, who discovered vitamin C: "Discovery consists of looking at the same thing as everyone else and thinking something different."[80]

Sensitive antennae. Homospatial thinking. Remote associations. These are all different manifestations of the same talent—fast and distant travel—that a creative mind displays during the minutes or hours of a psychological test. A creative mind's flashes of brilliance during such a test are a microcosm of human creativity, but the same talent creates the fireworks of major creations, in which knowledge accumulated during a life's journey becomes recombined. Such recombination becomes easier when our minds dream, play, or incubate, suspending the judgment—temporarily—that usually weeds out the inferior ideas that can ultimately become stepping-stones to perfection.

Mental recombination is as important to our creativity as DNA recombination is for biological evolution's exploration of adaptive landscapes. So too are play and various other means to reaching a state of mind that allows us to leap or drift through mental landscapes. And because landscapes are so central for all kinds of creation, the landscape perspective on human creativity can help us answer questions that matter to many of us: how to raise children, how to run innovative businesses, or how to turn entire nations into creative powerhouses.

Chapter 9

From Children to Civilizations

By the time South Korean students finish *suneung*, the national college admission exam, many have spent months sleeping fewer than six hours a night and studying up to thirteen hours a day, often in soul-crushing cram schools called *hagwons*—little wonder that they hurl their textbooks out the window after they are done. Exam day itself derails daily routines across the country: mothers fill the country's temples to pray, city officials add extra subway trains to avoid delays, air-traffic controllers reroute planes to avoid disturbing the English-listening part of the test, and policemen stand by on motorcycles to speed latecomers to the exam. Protesters even suspend demonstrations, such is the grip of the test on the public imagination.[1]

At least as important—and as harrowing to students—is the *gaokao*, its Chinese counterpart, taken by more than nine million Chinese students every year. Their preparation begins at age five, when Chinese "tiger mothers" begin to drill multiplication tables and proper syntax into them. Success means university and hopefully a better life, while failure can mean a life of manual labor. And where Chinese parents are hopeful that their kids will ascend the social ladder, many well-off US parents are afraid that their children will descend it. The costs for kids in pressure cooker high schools can be just as high. As one Palo Alto high school junior says: "We are not teenagers. We are lifeless bodies in a system that breeds competition."[2] With that kind of pressure, it is little surprise that both the *gaokao* and the rigors of elite US high schools have been linked to spades of teenage suicides.[3]

Children and parents like these live in an educational world governed by a simpleminded Darwinian precept as inescapable as gravity: hypercompetition is the only path to success. The more hyper the competition, the greater the success. In that pitiless world, depressed and suicidal youngsters are collateral damage, left behind in a race to the top where nothing matters except one product: nationally and internationally competitive students. The Asian system can clearly manufacture that product, at least according to international comparisons like the Programme for International Student Assessment (PISA), where Chinese and South Korean students routinely achieve top math, language, and science rankings.[4] That's why some educators and politicians view Asia's educational gulags with admiration.[5]

Charles Darwin—himself a middling student—might have recoiled at this dog-eats-dog view of education, even though it was arguably his theory that enshrined the value of competition in our minds. At the same time, he would have been hard-pressed to argue what's wrong with competition.

Nothing, it turns out. Competition is essential.

But it is not sufficient.

It is not sufficient to educate people who will live fulfilling, productive, and, most of all, creative lives. Their creativity is not a luxury. It is increasingly essential to building innovative organizations and a dynamic society. In a 2010 IBM survey, more than fifteen hundred CEOs in thirty-three industries cited creativity as the most important factor for business success.[6] And former US president Barack Obama said this at a 2010 conference: "Our single greatest asset is the innovation and ingenuity and creativity of the American people. It is essential to our prosperity and it will only become more so in this century."[7]

Although competition is often seen as essential to this creativity-based prosperity, competition alone cannot build it. Landscape thinking is helpful when trying to find out what *can* build it. This chapter describes how we can apply landscape thinking to the policies and politics that undergird a society to see which ones we should keep, which ones we should adopt, and which ones we should abandon to make our people, organizations, and nations as creative and innovative as they can be.

One of the greatest shortcomings of our present, hypercompetitive educational system is the punishing consequences it exacts for failure and the extreme emphasis it places on standardized tests. High-stakes mandatory tests like the Scholastic Aptitude Test (SAT) cause a creeping erosion of learning and teaching that starts in kindergarten and continues in grade school, where drilling in math and language displaces music, arts, and simple play.[8] Standardized high-stakes tests are not bad just because they steal time from true learning. They also homogenize education. That is especially insidious from a landscape perspective, as it eliminates the possibility of mental recombination not just within a single mind, but also among minds with a diverse set of knowledge and skills.

Homogenization was the intended effect of the predecessor to China's *gaokao* exam, the imperial *keju* exam, which was used to select top government and court officials for more than thirteen hundred years. With success rates as low as 2 percent, the *keju* required years or decades of preparation. Candidates were tested on subjects like civil law and taxation, but most importantly on the Confucian classics, which ingrain the values of order and obedience.[9] The *keju* not only helped build a meritocratic government, it also instilled precisely the values that keep an authoritarian emperor in power. "All heroes under the sun have fallen into my trap" is what seventh-century emperor Taizong observed upon watching successful *keju* candidates arrive at his court. Many of those who had failed the exam became teachers to *keju* candidates and perpetuated the Confucian message, whereas others had their rebellious streak rubbed out by years of studying Confucian classics. For more than a thousand years, the strategy that Tang Dynasty poet Zhao Gu called giving "white hair to all heroes" fed a homogeneous and obedient elite to the government.[10]

Chinese and other Asian governments today are aware of the damage that a system of hypercompetitive testing does to society.[11] Unfortunately, in China as in the West, children have become hostages to the system because opting out of a college admission test means gambling with a child's future. This situation persists, even though hard data proves that a test-driven education stymies creative thinking. Just consider Torrance test scores as one measure of creativity. By that measure, the creativity of 250,000 US children between kindergarten and twelfth grade has fallen steadily since 1990, even while IQ scores have been continually increasing. Although television and mobile devices play a role, they cannot take all of the blame. In US schools, the likely culprit is standardized testing and the escalating time taken for test prepping.[12] Sadly, Chinese kids are even worse off. One study asked 139 undergraduate students from two highly selective US and Chinese universities (Yale and Beijing) to produce artwork that was rated for creativity by nine judges, some of them Chinese, others American. Both the Chinese and American judges found that the artwork of the Chinese students was less creative than that of the Americans, regardless of the Americans' ethnic backgrounds. In addition to broader societal patterns, the authors faulted the Chinese education culture, where the stakes of testing are even higher than in the United States and where students have even less time for creative endeavors.[13]

The hypercompetitive education model ignores a core lesson from landscapes, namely that reaching peaks requires times of unfettered exploration during which selection and judgment are suspended. The youngest, preschool-age children especially learn much better by such exploration than by academic classroom teaching. A study of 343 children compared students who attended an academically oriented preschool program, where teachers instructed them in a classroom, with children in programs that allowed them to pursue "child-directed activities"— mostly just plain-old play.[14] By grade four, the students who had attended the academic preschool had lower grades than the kids who had just played. What is more, the academic children had missed out on the kind of exploration that is as important to becoming human as playing is to a great many other species.[15]

Even in older students immersed in the academic rigors of math, science, and grammar, creativity-building exercises remain important. Such exercises go beyond conventional art and music

classes and are easy to embed in any curriculum. A case in point
is the Private Eye Project created by Kerry Ruef, an American
educator who aims to enhance children's creativity by centering
on the kind of analogical and metaphorical thinking that creates
those remote connections so central for literature and science.[16]
Using tools as simple as a jeweler's loupe to observe objects like
seashells, insect wings, and dry leaves, its open-ended exercises
begin with deceptively simple questions, such as "What does it
remind you of?" A dry leaf will remind kids of a snake's scales, a
rotten bone, a beehive, a knitted blanket, a flag with holes, or
flaking skin. These similes can become the starting point of a
painting, a lesson in the physics of desiccation, or a story ("Yesterday I was a green flag, but now I am full of holes."). It is perhaps no coincidence that Ruef is also an award-winning poet.

While kindergarteners can already benefit from Private Eye,
the Berkeley-based program called Studio H requires the sustained attention of older children. Part of the regular curriculum
of the Realm Charter School in Berkeley, California, Studio H
was founded by architect Emily Pilloton and builds real-world
projects like farm stands, farmer's markets, or tiny houses for
homeless people. To build such a house, students would first
learn about architectural drawing, model building, and tool
wielding. They would then develop multiple designs—divergent
thinking—and whittle them down to a single one—convergent
thinking—which they would then build as a team.[17]

Studio H is sophisticated, but even simpler programs and a
small dose of play's suspended judgment can have powerful long-
term effects. In a Spanish study, eighty-six Spanish schoolchildren
aged ten or eleven were subdivided into two groups and followed throughout almost an entire school year. In the first group,

fifty-four children participated in exercises meant to enhance creativity. For one such exercise, students were paired up, and one student had to draw an animal. Her partner then had to start another drawing from a body part of that animal, such that the ears of an elephant might become the wings of a butterfly. In other exercises, students created imaginary advertisements, invented new names for familiar objects (*puree-launcher* for *spoon*), or conceived telephone conversations between unusual partners, such as a cow and a duck. In the second group, the remaining thirty-two children worked on art projects that were a normal part of the school curriculum, but not deliberately designed to enhance creativity. At the end of the year, children in both groups were tested on their ability to change from one line of thought to another, to create novel and unusual ideas, or to fashion images that two artists evaluated for originality. Not only did the children with creative playtime display greater creativity than the other group, but those children whose creativity scores had been the lowest before the experiment improved the most through playing.[18] Other studies also show that practicing creativity resembles learning a sport: anyone following the right exercises can learn to play tennis, even though few may end up playing it like Federer.[19]

But any amount of creativity training does not solve the problem of how to select the best candidates for success in college and in a profession. Fortunately, there are alternatives to the insidious standardized test.[20] Grade reports or teacher's letters may seem quaint, but they can work well. Case in point: a 2014 study of 123,000 US college students enrolled in thirty-three colleges that did not require standardized test scores for admission. It showed that good SAT scores did not predict the students' college grades nearly as well as good high school grades

did. Students who had tested well but had low high school grades
also had lower college grades and were less likely to graduate.[21]
Another alternative is assessment centers, which originated
when the US intelligence services needed to select spies during
World War II. To this day, corporate assessment centers evaluate the best candidates for skilled jobs in companies like AT&T
and General Electric. Candidates can spend up to several days
at such a center, where they perform individual tasks, undergo
interviews, and participate in group activities, which evaluate
not just hard, job-specific skills, but also motivation, teamwork
skills, and emotional intelligence.[22]

Grade reports and assessment centers allow merit-based selection while not automatically penalizing unusual curricula or
combinations of skills.[23] That is, they do not penalize diversity in
individuals or in groups. And diversity is landscape thinking's most
basic prescription for creativity. Human knowledge is so vast that
even with twenty years of schooling, a single mind can only absorb a minute fraction of it—the fraction that it can recombine.
By endowing different minds with different skills, a diverse education increases the chances that some of them will harbor the
right combination to solve tomorrow's hard problems. A society's
creative potential can unfold only if different schools share little
except a small set of core subjects, such as math and language, and
if students within a school can explore different combinations of
subjects—scientific and artistic, theoretical and practical, academic and vocational. An education driven by standardized testing
has the opposite effect: not only does it make students run up a hill
at maximum speed, it also makes them all run up the *same* hill.[24]

Diversity in education also requires that teachers have autonomy to teach diverse content and—just as important—that

students have autonomy to pursue idiosyncratic interests. The reason: intrinsic motivation. It's what psychologists call the desire to do things even when no extrinsic reward beckons. Its opposite, extrinsic motivation, can stifle creativity. One study, conducted by Harvard researcher Teresa Amabile with creative writers, demonstrated that even *thinking* about extrinsic rewards can be bad for creativity.[25] In the experiment, Amabile subdivided a cohort of writers into two groups. The first group had to complete a questionnaire on the intrinsic joy of creative writing, and the second on extrinsic rewards, like teacher approval, financial security, or a best-selling novel. Members of both groups were then asked to write short poems, which were judged by a dozen poets. According to the jury, the writers who had contemplated extrinsic rewards wrote worse poems.[26]

Object lessons in the power of intrinsic motivation are those exceptional creators who found their calling in childhood, such as Harvard biologist E.O. Wilson, who became a naturalist before age ten; chemistry Nobel laureate Linus Pauling, who started his first chemistry experiments at age thirteen; and astronomer Vera Rubin, who discovered her love for astronomy in her first decade.[27] Their inner drive could have been crushed not only by rigid curricula, but also by heavy-handed teachers, boring schools, stifling bureaucracies, mind-numbing memorization, and repetitive tasks.[28] The immensely quotable Einstein said it best: "It is…nothing short of a miracle that the modern methods of instruction have not yet entirely strangled the holy curiosity of inquiry."[29]

But nurturing the autonomy that preserves intrinsic motivation is easier said than done. Creative pupils often display traits like risk taking and impulsiveness that can be at odds with the responsibility and dependability that make a teacher's workday pleasurable—or merely bearable. And while most teachers will claim that they love to work with creative children, a string of psychological studies shows otherwise: they really love punctual, courteous, and conforming teacher pleasers with good grades and often stigmatize the energetic, rebellious, nonconformist, creative types as troublemakers.[30] But asking teachers to bear the sole responsibility for nurturing autonomous curiosity is really a bit much—who among us could fault a teacher for trying to keep a lid on thirty bouncing kids? Parents must do much of the work here. An important part of that work is choosing the right style of parenting, one that combines emotional support with academic involvement. The former encourages autonomy, while the latter prods students to work beyond the mandatory minimum and does not just push them but rather challenges them to push themselves.[31]

In sum, landscape thinking carries a simple and universal message about childhood education: cultivate diversity and enable autonomy. Such an education will endow individual minds with diverse skills, and it will endow different minds with different skills—recombination-ready. On the way to acquiring these skills, creativity training and playful learning will put to good use the time that test preparation wastes in today's hypercompetitive schools. And as we shall see next, the same principles apply in different guises to universities primed to educate tomorrow's creative elites.

S ometime in early 2009, I received an application to our Ph.D. program at the University of Zurich. The applicant did not come from a top university, his early college grades had been middling, and he had been out of academia for a few years. Normally, I would not take on a student like that, but the application was articulate, well researched, and carefully reasoned, and his references glowed with enthusiasm. So, I took a chance on Amit Gupta.[32]

I would not regret it. Over the four years of his dissertation research in my laboratory, we had many conversations where Amit proved to be a fountain of ideas for his own research and an enthusiastic collaborator whose creativity lifted everybody's boat. (He always balanced working hard with his passion as an amateur rock musician, which also helped explain his earlier grades.) And, four years later, he accomplished what other young scientists would kill for: the publication of the breakthrough results of his research in the most august scientific journal *Nature*.

Recognizing creative talent does not get easier from high school to grad school, but some things don't change: standardized tests like the Graduate Record Examination (GRE) do not help much. Amit would have fallen through the cracks in many graduate programs that impose test-score cutoffs or minimum grade-point averages. I myself might have tossed his application aside if years ascending a steep learning curve on selecting talent had not taught me otherwise.

My ascent on that curve began after I had left my home country of Austria—as a product of a solid but staid 1980s Germanic education—to attend graduate school in the United States. A few years later, Ph.D. in hand, I entered, like practically all my colleagues, the peripatetic life of a postdoctoral fellow, a journeyman researcher with a Ph.D. but no permanent job. That journey involved way-stops in both the United States and Europe before I settled into a faculty position at a US university and eventually moved back to Europe. My final destination was Zurich, Switzerland, fifteen years and six transatlantic moves after I had first left the Old World.[33]

In both the United States and Europe, I served on numerous admission committees for master's and Ph.D. programs that selected the best from among hundreds of applicants. Time after time, we would interview Olympian test-takers, many of them from countries where rote memorization and test scores were the be-all and end-all. And time after time, we would be disappointed. To be sure, their memorization muscles were powerful, but their schooling had prepared them poorly for the real life of a scientist, where rote memorization does not help one imagine a great—or even just a good—experiment. Sadly, some of these mental gymnasts did not even understand the questions behind their modest college research projects. Like soldiers following a general, they had just blindly executed the commands of a professor.

As a university educator and researcher, I also received other field lessons in the perils of rote training and conformism. One of them came during a half-year research stay at a government research institute in Singapore. Most doctoral and postdoctoral researchers I met there had survived the rigors of Singapore's schools and universities, which are well reputed in Asia. Aside from being exquisitely polite, the researchers were technically highly competent and excelled at analyzing mountains of data. However, their training had not prepared them to take the essential step toward a creative career: leave behind the authority of your professor and bushwhack your own trail, following only the lead of the questions that excite you. It is sad to think what could have become of these bright and hard-working kids, whose research had a robotic quality and lacked creative spark, if their education had nurtured that spark.

But along my path I also frequently stumbled upon diamonds in the rough, like Amit, candidates with less than stellar scores whose minds shone brilliantly. Once these diamonds had been cut by years of scientific research, they sparkled through their experiments and theories, which went on to command international attention. Students like these were often neither the most single-minded nor the straight-A students. Their college grades might have taken a dip when they played in a band, pursued competitive sports, or traveled around the world, but in mysterious ways all those distractions helped make them more passionate and creative in their work. More than half a century ago, Spanish Nobel laureate Santiago Ramón y Cajal—a father of modern neuroscience—already knew this when he recommended that one choose "students...who being endowed with an abundance of restless imagination, spend their energy in the pursuit of literature, art, philosophy, and all the recreations of mind and body. To him who observes them from afar, it appears as though they are scattering and dissipating their energies, while in reality, they are channeling and strengthening them."[34]

Observing the rise of these scientists also taught me a few lessons about building and running a team of creative researchers—currently some twenty members strong—that maps evolution's adaptive landscapes.

Rule one I already mentioned. Before you hire a person, really get to know her. No test score is a substitute for lengthy interviews, writing samples, presentations, and conversations with references.

Rule two is about autonomy. If a researcher has the skills and the passion, give him the freedom to pursue it. That's why members of my team usually develop their own research projects. They usually come up with multiple options—divergent thinking—which we whittle down to the one that is most original, feasible, and exciting.

And we must be patient. Many great projects are difficult. They navigate a complex landscape, and, like for Dante, the road to heaven often leads through multiple hells—months of fits and starts and failed experiments. If fear of unemployment lurked behind every failure, a researcher's imagination would stop leaping and bounding and would gamely climb up the nearest hill to stay there.

A final rule, this one also familiar: diversity pays. I don't just mean the diversity of skills in any one person's head (that too), but also the diversity of a team whose members have different backgrounds. Even though we are all interested in biological evolution, my team has hosted not just biologists, but also computer scientists, chemists, physicists, and mathematicians. Each has their own vantage point on the landscape of human knowledge.

This kind of diversity is essential to thriving research collaborations, where success has as many parents as a team has members. This is not just one man's opinion. A 2007 study analyzing almost twenty million scientific works published over a half-century demonstrated that scientific teamwork has been becoming inexorably more important than solitary work. The reasons are obvious in science and engineering, where research has become ever more costly and may require equipment from multiple laboratories. But teams replace individuals also in the social sciences, the arts, and the humanities. And even in mathematics, historically the purview of the lone genius, much successful research is now performed in teams. Indeed, the most influential research publications—those cited more than a thousand times—are six times more likely to have been written by teams than by solitary scientists.[35]

In sum, academic research, like childhood education, is enhanced by autonomy, diversity, failure tolerance, and the recombination of knowledge. Landscape thinking can not only help explain why, it can also help neutralize two threats that loom over today's fundamental research.

The first comes from how traditional universities are organized, where researchers are neatly boxed up in departments—economics, physics, art history, biology. When collaborations span different departments, these very boxes can hinder the successful recombination of skills and ideas. Fortunately, university researchers fed up with academiosclerosis have options—small organizations like the Institute of Higher Studies in Berlin, Germany, or the Santa Fe Institute in Santa Fe, New Mexico. These are institutions that exist for one purpose: recombining knowledge. They may host a few or no permanent resident scientists, but rather numerous visiting scholars from all disciplines, who stop by for anywhere from a few days to a few years to exchange ideas.

I have been visiting the Santa Fe Institute regularly for almost twenty years, and on these visits, I never know who I will meet—writers, physicists, educators, archeologists, or biologists. But I do know that kicking around ideas with them will be as stimulating as LEGO is to children. Many of their ideas linger in my mind for weeks, eventually germinating into new research projects and discipline-crossing books—like this one.

The tricks by which small institutions can accelerate recombination are hard to emulate at larger universities. One of them is smallness itself, which avoids anonymity. By constantly bumping into each other, visitors are bound to engage. Then there is a semi-secluded location, preferably outside a city center, which prevents people from wandering off for lunch or coffee—they talk to each other instead. Shared meals served on-site achieve the same effect. And finally there is the physical space. Large and comfortable communal areas stimulate scientists to exchange ideas, contrasting with the small and shared offices given to even the most distinguished visitors. (Try enforcing that at a university.)

While institutes like these are no substitute for the teaching portfolio and billion-dollar infrastructure of universities, they are models for effective recombinators of the future. They already leave a footprint far beyond their size and have produced hugely influential collaborations in areas as different as sociology and biology.[36]

But while such institutes can combat academic provincialism, they are powerless against the second, even greater danger for Western science: to become a victim of the spectacular success it has seen since World War II.

In the United States, the foundations of this success were laid by visionaries like Vannevar Bush, science advisor to president Franklin Delano Roosevelt. Bush wrote a 1945 document entitled "Science: The Endless Frontier" that helped transform the United States from a scientific backwater to a global discovery engine. In this document, Bush argued that the government should fund basic research that is unconcerned with immediate application—the kind that led to the discovery of antibiotics. He envisioned "affording the prepared mind" of the researcher "complete freedom for the exercise of initiative."[37]

His vision helped create funding agencies such as the National Science Foundation, which has supported the work of countless US academics, including more than two hundred Nobel laureates.[38] Yet Vannevar Bush would be dismayed to see what is becoming of the freedom that he envisioned.

Just like education, scientific research has become hypercompetitive. A primitive social Darwinism would welcome that development, but the landscape perspective exposes its dangers. In today's US universities, more than one hundred thousand biomedical faculty help manufacture sixteen thousand new Ph.D. or M.D. researchers every year. These researchers join an army of more than thirty thousand postdocs, who all scramble for scarce academic positions.[39] And for the select few who make it, the hassles are just beginning. They enter a competition for funding so intense that they have little time for the very research they excel at. Instead, they write proposal after multipage proposal, which compete with hundreds of other proposals at funding agencies like the National Science Foundation, where success rates in some areas of science have dropped to below 5 percent. In other words, the average young university scientist—already one of a select few with an academic job—has to write twenty proposals before being able to do what he is good at. And without that funding, his career will end before it has really begun.[40]

The experts reviewing those proposals are so overwhelmed that they are looking for any excuse to reject a proposal they read. Young scientists know that well, and they also know that expert panels are filled with older scientists who tend to be conservative. An original and unusual proposal can be just the excuse a conservative expert is waiting for. The natural reaction: stick to safe research. Such research is like a cookie-cutter suburban home compared to a radically innovative building like the Sydney Opera House: any reasonably competent builder could assemble it. In other words, such research is guaranteed to succeed, but it will not produce any breakthroughs. Hypercompetition turns visionary architects into uninspired craftsmen.

Whether experts recommend a proposal for funding also depends on a scientist's track record. To evaluate that record one must read an applicant's work, but given the number of applicants, that's completely out of the question. So the seduction of using a single number akin to a student's test score is irresistible. In science, that number is the number of citations a researcher's work has received. Citations are great for studying broad historical patterns, but they are dangerous when evaluating the work of young scientists. They can be especially misleading when evaluating groundbreaking work, which may take years to become recognized and cited.[41] By the time the world has caught up, the young scientist may be driving a taxi.

These are some symptoms of academic hypercompetition, which reduces diversity among scientists just as it does among pupils. The resulting homogenization has been quantified in economics research in the United Kingdom, where the government is granting or denying research funding to entire university departments based on the collective impact of the departments' researchers. Between 1992 and 2014, economics funding became concentrated at fewer universities, which therefore attract a greater percentage of students. They publish in fewer journals, research more mainstream subjects, and teach orthodox, mainstream economics.[42]

The problem with this trend: most scientific revolutions originate far from the mainstream. When everybody scrambles up the same hill, a landscape of knowledge remains unexplored, and discoveries like antibiotics or the DNA double helix go unmade. Hypercompetition produces scientific monoculture, just like high-stakes testing does in schools.

But what to do?

The post-War golden age of increasing research funding is a distant memory, so universities will need to curtail the number of academics they produce.[43] That's one part of the solution. Another part emerges from the landscape view and from the importance it places on allowing the occasional failure: give young and promising academic researchers a modest amount of research funds, independent of any competitions they win—not much, but enough to pursue unusual ideas.[44] Don't take that money away if they fail, so they can try another idea. If they succeed, let them compete to obtain more funds to progress more quickly.

Safe but modest funding not only takes the sting out of failure and helps young researchers cross the personal research hell that can precede breakthrough discoveries. It also steers them away from the latest research fashions—the kind they need to pursue for publishing in fashionable journals. It helps young scientists remain explorers, regardless of whether they end up on the Discovery channel.

This experiment is ongoing in some European countries. Among them is Switzerland, where providing such safe funding is standard practice.[45] Being a tiny country, Switzerland cannot produce the same quantity of research output as the United States. But, remarkably, if one tallies its research products per capita, Switzerland is either on a par with or ahead of leading nations—including the United States—in the number of publications, their influence in several fields, and the number of Nobel Prize winners. There are multiple reasons for this, but among them is that the best Swiss universities provide academic researchers the means—via limited safe funding—to traverse the valleys that lead to the highest peaks. The message is one for politicians everywhere: dialing back on relentless competition is not synonymous with wasting taxpayer money. Done right, it can be a recipe for innovation.[46]

Most businesses that develop new products cannot take the long view on exploring creative landscapes that Switzerland's academia has taken. They need to transform ideas into products within months or at most a few years. That's why most of them do not develop radically new technologies, which might take decades to commercialize. Rather, their research and development tweaks or optimizes existing products. In other words, their R&D is mostly D and much less R, climbing nearby hills rather than distant mountains.[47]

Perhaps for this very reason, surveys of business leaders reveal that they value university research very highly, not just for the future employees it trains, but also for the diversity of knowledge it creates. In the landscape view, university and industry have a symbiotic relationship. The first jumps to distant places, the other ascends from there in smaller steps.[48]

But some companies have broken this traditional relationship. They spend decades commercializing discoveries made in academia, such as graphene, genetic engineering, or computer-controlled machines. Others operate their own engines of discovery.[49] Such unusual companies hold especially valuable lessons for how to run creative businesses.

Historically, among the first—and certainly the best known—was AT&T and its research laboratory, Bell Labs, which mothered innovations as groundbreaking as transistors, solar cells, lasers, and fiber optics. When its star rose half a century ago, landscape thinking had not yet left biology, but the principles on which Mervin Kelly ran it—he was one of Bell Labs' chief architects—could be taken straight from Landscapes 101.

The first is diversity. Kelly paired physicists with electrical engineers, thinkers with doers, scientific celebrities with rookies, all in the same space, to create the remote associations we now know are essential for the highest creativity. The second is freedom and time—lots of them. Years could pass before some projects would bear fruit, which was only possible because no product had to be rushed to market. Time buys the luxury of failure.

These principles are simple, so why doesn't everybody follow them? In the words of physicist Phil Anderson, whose Nobel Prize was itself a product of Bell Labs: "Never underestimate the importance of money." Bell Labs was sponsored by a giant telephone monopoly with deep pockets.[50] Only some of today's wealthiest companies like Google can afford the patience needed to innovate.[51]

Such patience is important in determining whether a company's research breaks new ground. Teresa Amabile and her team have studied multiple companies where creativity affects the bottom line and have learned valuable lessons about what works for them and what doesn't. One of their lessons is about patience. Many of us feel that tight deadlines spur our creativity, but according to Amabile's research, extreme pressure puts the brakes on creative thinking. Workers with sufficient time to solve a problem will usually find more original solutions than those under adrenaline-raising pressure. What is more, once the stressed-out workers have met a tight deadline, they often experience a creativity "hangover" in which their minds produce ho-hum ideas for days.[52]

Encouraging creativity also requires breaks and vacation. When Socrates warned us of the "barrenness of a busy life" more than two thousand years ago, he may have already realized that important ideas need time for incubation. And today's psychological research shows that creative people, even though they sometimes work themselves into exhaustion, also take more time off than others.[53] The positive effects of letting a mind roam free are even measurable. For example, in the accounting company Ernst and Young, every ten additional hours of vacation help improve an employee's performance review by 8 percent.[54]

A third lesson is that a creative team's members must differ in skills, interests, and perspective. The reasons are obvious from landscape thinking—diversity allows the recombination behind remote association. Unfortunately, many managers are not ready for that lesson, perhaps for the same reason as teachers prefer docile over rebellious pupils. They assemble homogeneous teams whose members are like-minded, eager to work together, and therefore have high morale. But Amabile's work shows that homogeneous teams are creativity killers.[55] They ascend minor hills and create mediocre solutions.

Building a creative team of adults is certainly not easier then herding creative children. It requires vision, intuition about what motivates people, self-confidence in the face of conflicts, and a tolerance of large egos. But companies that are up to the task can become creative powerhouses, like design company IDEO, best known for the first Apple mouse. Its leaders excel at building diverse teams, and these teams in turn have not just created products as different as body-hugging office chairs and wearable breast pumps, they have also reaped hundreds of design awards in the process.[56]

A fourth lesson is that human creativity requires the freedom to explore a landscape, the same kind of freedom that genetic drift allows in evolution's adaptive landscapes. Managers who subject every idea to layers of review or early criticism are poison to creativity. Selection is important, but not early on. And even managers who reward originality can go wrong: praise is helpful, but cash prizes are not. Not only do they make employees feel manipulated by management, but they also turn creators into mercenaries. In other words, they eliminate the intrinsic motivation—creating for the fun of it—that is so important for exploration.[57]

Relatedly, business creativity also needs autonomy, just like Bell Labs pioneer Mervin Kelly realized. Developers may have to meet strategic goals, but how to get there should be left up to them. Without such autonomy, Bell Labs' scientists would not have discovered the transistor, a breakthrough innovation that could not have been made to order.[58] Such autonomy also buys the freedom to commit errors, dead ends from which a creative journey has to retrace its steps. While traditional businesses invest top dollar to prevent such errors, psychologists who study organizations declare that investment to be wasted. Rather, errors need to be managed to minimize their negative consequences and to amplify their positive consequences—learning and innovation.[59] After all, they are not only unavoidable, but they can also lead to serendipitous discoveries, like Teflon, penicillin, and vulcanized rubber.[60]

All this goes to show that creative businesses run on the same fuel that helps educate creative citizens and accelerate basic research. That's just natural. After all, they also explore the landscapes of creation. And thinking in terms of these landscapes provides a single framework to explain the successful practices of creative businesses.

If creativity is a national asset, as Barack Obama posited in 2010, then governments should aim to grow this asset. Landscape thinking can help us see how the right laws and regulations can achieve that.

Let's first talk about diversity. Part of it is a numbers game—the more people that explore a landscape, and the more diverse their skills are, the more of them will succeed in finding new peaks. The industrial revolution, for example, became possible only when science and engineering were no longer the privilege of a few aristocratic scientists and began to be practiced by many craftsmen who needed to make a living. Their trials—and errors—created a huge burst of technical innovations.[61]

But numbers are not everything. A society's bouquet of skills can be a riot of wildflowers or a bunch of tulips. Laws that regulate schools and universities influence which it is going to be. Most Chinese schools and universities, for example, are not just branches of the government. Their curricula are also controlled by the government, as are those of private schools. Only international schools are beyond the government's reach, and they provide a unique opportunity at cultural recombination. Alas, they are off-limits to Chinese citizens.[62] (Western education is more diverse, but to the extent it will continue to teach to the test, its future will also spell monoculture.)

Even when a nation does not grow enough diversity at home, it could import such diversity through skilled migrant workers. Countries that suppress recombination through migration do so at their peril. A case in point is Japan, a notoriously homogeneous society where foreigners account for only 1.5 percent of the population. Japanese companies complain about useless university graduates without international experience, but they prefer to hire precisely those homegrown graduates lacking unorthodox training and foreign experience. Even more important, Japan is not a welcoming place to foreign talents: a measly 3 percent of students in Japan's universities are foreigners, and their numbers are even falling. Moreover, fewer than 4 percent of university professors are foreign. It is perhaps no coincidence that many Japanese businesses struggle to compete abroad, and only two Japanese universities rank in the top 100 worldwide.[63]

Other countries do much better, with 25 percent of science and engineering graduate students being foreign in the United States, and as many as 40 percent in countries like the United

Kingdom and Switzerland. And this diversity matters. It matters even in the microcosm of creativity tests, where participants who have lived in different countries score better. As for the macrocosm of major creative works, we just need to remind ourselves that 44 percent of US Americans with outstanding contributions to science and society are recent immigrants.[64]

But room for improvement exists even in countries that are permeable to immigrants. For example, foreign students are excluded from a major source of research funding for US graduate students—training grants made to universities by government agencies such as the National Institutes of Health. The smart policy—advocated by academic leaders—is to include foreign students.[65] And industry leaders like Microsoft's Bill Gates and Facebook's Mark Zuckerberg have complained for years that US universities train countless foreign math and science Ph.D. students, only to send 40 percent of them back home because of draconian visa policies. All the while, US technology companies hungry for their skills are left starving.[66]

Tech is not the only sector that benefits from the diversity created by migration. An example with hard data is the fashion industry, where sales and profits of fashion houses are driven by one creative product, the fashion collection. Single individuals wield enormous influence on this product—and on the financial bottom line of their fashion house. These influencers are creative directors like Karl Lagerfeld, Giorgio Armani, or Tom Ford, who define the creative vision of a collection. Many of them have extensive international experience or multicultural backgrounds. Karl Lagerfeld, for example, has a Swedish father and a German mother and commutes between France and Italy for his work. Every fashion season, the collections that he and other creative directors produce are rated for their creativity. The raters are the very industrial buyers that determine whether a collection makes it into stores. These ratings, which are published in the prominent trade magazine *Journal du Textile*, can help link a director's international experience to both a collection's creativity and its profitability. A 2015 study did just that for collections of 270 fashion houses rated by more than sixty buyers during twenty-one fashion seasons. And it produced a clear answer: the more years a director spent living outside his home country, the more creative his collections were. The study sheds new light on Lagerfeld's ability to elevate not one but two companies, Chanel and Fendi, into the top ten of global luxury brands.[67] But, more important, it underscores the importance of making country borders permeable to skilled labor.

Advocates of strict immigration policies could learn not just from such modern examples. They could also use a history lesson. What I have in mind are not just the life histories of the great artists we have encountered, painters like Paul Gauguin and Raphael, whose itinerant lives helped them create new styles of painting. I mean a lesson about the creative ebbs and flows of entire civilizations, like those studied by creativity researcher Dean Simonton, who analyzed how the creative output of five thousand Western writers, philosophers, scientists, and composers fluctuated over two and a half millennia, from 700 BC to 1800 AD.

Like others before him, Simonton found that most eminent creators emerged in times of political fragmentation, such as Renaissance Italy and ancient Greece, when many independent states co-existed.[68] Such fragmentation enhanced a region's cultural diversity and promoted the recombination of ideas.

In contrast to fragmented societies, large empires like the Roman and Ottoman empires brought forth fewer influential thinkers. But even in such empires, diverse minority cultures with divergent interests persisted. They made their presence felt periodically through political upheavals like revolts and rebellions. And just like a burst of wildflowers follows a desert rain, a bloom of creative work often followed about a generation after such upheavals.[69]

Simonton also studied the creative history of Japan between 580 AD and 1939 AD. During this time, spanning more than a thousand years, Japan alternated between periods of openness to foreign influence and periods of near complete isolation. Combing through historical records about eminent creators in areas as different as literature, sculpture, medicine, and philosophy, Simonton tallied the numbers of these creators at any one time. He also recorded foreign influences on Japanese culture, such as how many immigrants rose to eminence, how many creators went abroad to study, and how many admired foreign ideas. And, sure enough, shortly after a period where such foreign influences increased, Japan brought forth more individuals with notable and outsized contributions to Japanese culture.[70]

These lessons from Japan's history apply worldwide and to this day, at least for science. In a 2017 study of 2.5 million scientific publications, those countries that produced the most impactful publications were also the most open to immigrating scientists, to the mobility of their homegrown researchers, and to international scientific collaborations.[71]

But enough about diversity and recombination on a national and global scale. What about failure?

A nation's attitude toward failure is written in its customs, regulations, and laws, like those that govern businesses. Among them are bankruptcy laws. In most European nations, draconian bankruptcy laws punish creative entrepreneurs with failed start-up companies and can turn them into lifelong financial cripples. Add to that the social disgrace of business failure, and you have a foolproof recipe for stifling business innovation.

The tacit assumption behind punishing bankruptcy laws is this: business failure results from incompetence or irresponsibility. But in the United States, with its many start-up companies, the statistics tell another story. Even entrepreneurs with a previously successful start-up company have a less than one in three chance to succeed in their next business venture.[72] Much like success in biological evolution, business success is a lottery, and that holds also for the best and the brightest. That's why entrepreneurial giants like Nolan Bushnell, founder of the hugely successful 1970s Atari brand of video games, later produce major flops, forgotten companies like uWink and PlayNet.[73] It's another example of the hit-or-miss nature of creativity.

To enhance rather than suppress such creativity, the sting needs to be taken out of failure, and some societies are making progress in this area. The Silicon Valley mantra of "fail fast, fail often" speaks to this progress, as do increasingly popular social events where entrepreneurs share their experience with failure and celebrate it as a learning opportunity. They include the wildly successful FailCon conferences and the saltily named Fuck Up Night, a social event that originated in Mexico City. Within a few years, Fuck Up Night spread to more than two hundred cities in seventy-five countries.[74]

Public forums like these weaken the social sting of failure. The financial sting can be taken care of by a government's way of forgiveness: lenient bankruptcy laws. Here, the United States has also been doing well, mainly thanks to its Chapter 7 bankruptcy, which protects failed entrepreneurs from paying back most debt and allows them to keep some assets, such as a home.[75] A 2003 study led by San Diego economist Michelle J. White asked whether legal leniency like this can indeed lift the entrepreneurial spirit. The study examined entrepreneurs in ninety-eight thousand families across the United States. What made the study possible is that different US states do not treat bankrupt individuals equally. Some states, like Texas, allow for a large "homestead exemption," as lawyers call it. In these states, a failed entrepreneur can keep all or most home equity, whereas other states, like Maryland, allow her to keep little or none of it. In theory, a large homestead exemption should be good for business. It reduces start-up risk because it protects more of a family's assets from failure.

And this bit of economic theory works: families who own their home were 35 percent more likely to own a business if they lived in states where the homestead exemption is high.

Unfortunately, forgiveness is not free. In 2004, nine times more people filed for bankruptcy in the United States than in Britain, and to the chagrin of lenders their number has increased five-fold since 1981. That chagrin materialized itself in a 2005 tightening of the legal thumbscrews for US debtors. It essentially reduced the size of the Chapter 7 safety net such that now fewer individuals qualify for Chapter 7. Landscape thinking helps us predict how this long-term experiment will affect business innovation.[76]

Just as essential for national creativity as failure tolerance and recombination is the autonomy of individuals. That's a simple consequence of creativity's Darwinian aspect. Because we are blind to the best solution for hard problems, it pays to have many people tackle such problems. And if we let each institution, business, or individual explore a solution landscape in a different direction, we increase the chances of stumbling upon good solutions.

But while lawmakers can turn the dials of legal forgiveness by rewriting bankruptcy law, enhancing the autonomy of explorers can require more profound—not to say daunting—social changes.

First, individual autonomy directly contradicts some forms of government. Most authoritarian governments would be suicidal to grant it in even small measure, lest its people get a dangerous whiff of the freedoms they could have. In their book *Why Nations Fail*, economists Daron Acemoglu and James Robinson identify what has united nations—from the Neolithic to the modern world—that function well: a government and institutions where individuals are not at the mercy of a king, feudal lord, or dictator, but where they can shape their own fate. Because the same autonomy also helps creativity, a society's creative potential will grow with the number of people endowed with it. (And that potential will erode when a society's middle-class does—for example through income inequality.[77])

A second obstacle to autonomy is even more formidable than the form of government. It is a bedrock of social values that can discourage taking the road less traveled.

Students of civilization have long argued that humans in Eastern and Western cultures develop two different conceptions of self.

The West values an *independent* self that expresses its own needs and develops its own potential. In contrast, the East values an *interdependent* self that is focused on how to serve a larger whole—family, community, and nation—a value epitomized in the proverb "the nail that sticks out gets hammered down."[78] The roots of the Western perspective lie in the defense of civil liberties and personal freedom by eighteenth-century liberalism.[79] By comparison, those of the Eastern view are ancient, going back all the way to the emphasis Confucius placed on social harmony some two millennia ago.

Schools imprint these differences onto children. Western schools aim to develop and empower the individual, whereas Eastern education is more about socialization—teaching the individual to fit in and serve the collective. That's why the Imperial *keju* exam was more than just an emperor's way to neuter challengers, and the *gaokao* test is more than just a tool for meritocracy. They also sustain social harmony, which is easier to achieve when people think alike.[80]

These millennia-old cultural differences help explain why well-known Chinese inventions like the printing press, compass, or gunpowder transformed the West but made barely a dent in the East. Where stability, order, and harmony are cherished, disruptive innovations cannot be.[81]

In other words, if a standardized education is to blame for the less creative artwork put out by Chinese students, as shown by the Yale–Beijing study I mentioned earlier, that education is itself a symptom rather than the disease.[82] And the disease is not endemic to China. Twenty years before the Yale–Beijing study, researchers at Tel Aviv University evaluated the creativity of ninety children from the United States and the Soviet Union and found that the Soviet children scored lower on creativity tests.[83] Soviet schools were not known for Chinese-style hyper-competition, but Soviet society shared another feature with the Far East: they were collectivist and valued conformity, although for a different reason—communism.

But what about the one great advantage of collectivism, superior teamwork? That advantage is real, but remember that creative teams do not just divide labor among worker bees. They exist to recombine ideas, and that is where diversity is crucial: you cannot recombine ideas that no one has thought of.[84]

Homogeneity is not the only by-product of collectivism. Another one is what Kai-Ming Cheng, chair of education at the University of Hong Kong, calls "the extraordinary significance that extrinsic motivation plays in student learning."[85] Student achievement in a collectivist society is often driven by the expectations of others and rewarded by their approval rather than by inner drive.

Unfortunately, as we heard earlier, extrinsic motivation does not drive creative achievement. Intrinsic motivation does.[86] And the landscape perspective helps explain why: an explorer must blaze his own trail because nobody can know where solutions to hard problems might be found. An interdependent self-suppresses, whereas an independent self-supplies the courage to be different. That's why autonomy is so central for creativity.

From the microcosm of a child's mind to the macrocosm of civilizations, the same principles appear in different guises. To help recombine and associate remote ideas, single minds can acquire diverse educations, whereas nations can build networks of autonomous schools and facilitate the migration of skilled workers. To tolerate failure, weaken the tyranny of selection, and suspend judgment, nature has endowed individual minds with priceless abilities like those of playing and dreaming, and governments can endow an entire nation with forgiving bankruptcy laws. On the surface, Chapter 7 bankruptcy has nothing in common with a playing child, but the landscape perspective reveals their deep commonalities. And it reveals the hallmarks of creative societies: they treasure diversity, tolerate failure, and shelter the autonomy of individuals.

Epilogue

More than Metaphors

When physicists like Max Planck and Louis de Broglie developed quantum theory, they also discovered that vibrating strings are more than superficial metaphors for glowing atoms. In the century since then, another new kind of science has been emerging, this one fed by rivulets from disciplines as different as molecular biology and psychology. It is a science of the creative processes that unfold all around us, from chemistry to culture. And just as the idea of vibration ties together quantum theory and acoustics, optics and cosmology, the concept of a landscape ties together creative processes from chemistry to culture.

Landscape exploration is more than just a metaphor for creation. We know this for two reasons.

First, in biology we can now map these landscapes in deep molecular detail. They help us understand the evolutionary paths of proteins like beta-lactamase toward greater antibiotic resistance. And well-mapped adaptive landscapes can do more than explain creative evolution. They may even help future scientists predict it, just like a detailed map can show a mountaineer where the highest peaks are and how to get there.[1]

The second reason is that many acts of creation are acts of problem solving. A gleaming quartz crystal embodies a solution to the problem of finding a stable arrangement of silicon and oxygen atoms. A metabolic enzyme breaking down glucose has solved the problem of harvesting energy from carbon bonds. An ammonite has solved the problem of swimming with minimal drag resistance within the confines of its spiral architecture. And creative machines can use evolution's problem-solving strategies to invent new technology and compose delightful music.

Today we know that the robotic hill climbing of natural selection is a poor strategy for solving hard problems. Conquering a complex landscape needs autonomous explorers, be they organisms with DNA mutations or human trailblazers, who take off in different directions to create diverse solutions. It needs mechanisms like genetic recombination and remote association that promise access to distant peaks. And it needs mechanisms like exploratory play and genetic drift that can descend a landscape's many valleys and create poor solutions that become stepping-stones for better ones.

These mechanisms are at work from molecules to humans, and landscape thinking can help explain why phenomena as different as strong genetic drift and lenient bankruptcy laws serve similar roles in the creative process. Not only that, but landscape thinking also can help enhance human creativity, and it can do so for individuals and entire nations alike.

The key is balance. Harsh selection must be balanced with tolerance of failure, rigor with playfulness, convergent with divergent thinking, authority with autonomy, mindfulness with mind-wandering, educational depth with breadth, small steps with giant leaps. Compared to Darwinism 1.0, that insight is already revolutionary.

Unfortunately, we still know little about where this balance lies. This is even true for the creativity we can observe and control best, like that of bacteria evolving in my Zurich laboratory. For example, we do not know the right balance between the small steps of DNA mutations and the long leaps of DNA recombination that will teach a bacterium to survive a toxic molecule or a viral parasite. The genetic algorithms of computer scientists, which simulate evolving populations with tunable amounts of mutation and recombination, tell us that there may not even be a general answer. The right balance may depend on the problem to be solved.

Finding that balance for human creativity is a job for future generations, but in a world tilted far toward competition, some prescriptions are easy. Creativity-enhancing programs will move the scale in the right direction for children in a hypercompetitive school system. More-lenient bankruptcy laws will do the same for business innovation in countries that purge failed entrepreneurs, and migration will do the same in the least diverse societies. After more than a century of a simpleminded Darwinism, it will take a long time before the scales tilt too far the other way.

Landscape thinking also harbors some painful truths. The most obvious one is that failure is unavoidable. Biological evolution is blind, and so are we. In other words, creativity will always be inefficient. Biological evolution is inefficient because it eliminates the vast majority of new mutants. Fundamental research is inefficient because it must plant numerous seedlings to harvest a few luscious fruits. Business innovation is inefficient because it is littered with failed start-up companies. The inevitability of failure holds a lesson for those politicians who aim to eliminate all wasteful research: their efforts will achieve little more than to destroy a society's creative potential.

Sadly, inevitable failures also mean that there are no reassurances for the parents among us who worry that our children will take a dead-end path on their creative journeys. That's another reason why second chances are so important. To the extent that we can learn to tolerate failure, not just in playing children, but also in the more momentous experiments of scientists, strategies of companies, and policies of nations, we will approach our full potential to create a world of our choosing. The thirteenth-century theologian Thomas Aquinas was onto something when he wrote that God created the world in play.

Acknowledgments

I would like to thank the members of my research team at the University of Zurich for the many scientific discussions that have shaped my thinking about adaptive landscapes over the years. I am also grateful for the continued support of the Santa Fe Institute. Countless conversations with my collaborators at the Institute, as well as with resident and visiting scientists over the years, greatly helped expand my horizons beyond the realm of biology and into the social sciences, engineering, and the arts. This book would have been impossible without these conversations. Jeff Alexander provided much appreciated early structural advice on the manuscript. Thanks also go to T.J. Kelleher and Melissa Veronesi for their incisive editorial work. David Young Kim provided useful source material on artistic journeys. Lukas Keller, Melanie Mitchell, Carel van Schaik, and Dean Simonton provided feedback on parts or all of the manuscript. I have followed most but not all of their suggestions, and the book may be worse for it where I did not.

My agent, Lisa Adams, has been unfailingly professional and patient in addressing many not just contractual but also strategic and editorial questions. Last, but not least, I would also like to thank the editorial team at Basic Books for birthing the final product.

Notes

Prologue

1. Appropriateness is no less important than originality: "Forty-one" is a highly original response to the question about the sum of two plus two, but it is not appropriate. Psychologists also make further distinctions, such as that between the creativity of a product (the Mona Lisa, Beethoven's symphonies, or General MacArthur's battle plan) and creativity as a personality trait, really a disposition, found in highly creative people like Mozart or Einstein to bring forth creative products. See page 152 of Eysenck (1993). For the purpose of this book, a product-centered definition of creativity is most appropriate.

2. See page 33 of Kubler (1962).

3. Although, as I mention in Chapter 1, Wright's own experiments used guinea pigs.

4. It is worth keeping in mind that in the realm of biology natural selection is not synonymous with competition. For example, under fertility selection, where organisms in a population differ in the number of offspring they have, perhaps through their genetic makeup and not through any limitation in resources, more fertile lineages may come to dominate a population without any need for competition. If I juxtapose competition and selection throughout, it's because competition is perhaps the closest analogue to selection in the human realm.

5. See page 282 of von Helmholtz (1908). This edited volume is from 1908, but the passage in question stems from a lecture that von Helmholtz gave in 1891.

Chapter 1: The Cartography of Evolution

1. See page 38 of Clark (2013).

2. See Chapters 1–3 of Clark (2013).

3. See page 100 of Clark (2013).

4. See page 70 of Clark (2013).

5. See Darwin (1859).

6. See Vol. 1, Chapter IX of Darwin (1868).

7. For a concise summary of pertinent evidence and a spirited defense of Darwin's theory see Coyne (2005).

8. The term *survival of the fittest* was coined in 1864 by Herbert Spencer and adopted by Darwin a few years later in his fifth edition of *Origin*.

9. See Kettlewell (1973) and Majerus (1998). Note that these experiments were performed long after Darwin's time. See also the blog entry "The Peppered Moth Story Is Solid" by evolutionary biologist Jerry Coyne on his blog "Why Evolution Is True" at http://whyevolutionistrue .wordpress.com/2012/02/10/the-peppered-moth-story-is-solid/.

10. Haldane (1924).

11. Haldane called the difference in fitness between two organisms a *selection coefficient*, a term still used in textbooks almost one hundred years later.

12. See Chapters 5 and 9 of Provine (1986) as well as Wright (1978). By complex interactions I specifically mean that genes interact non-additively, non-linearly, or (in genetic jargon) *epistatically* when bringing forth a phenotype.

13. Wright (1932). An earlier predecessor was proposed by the French scientist Armand Janet in 1895, but it lacked the genetic component essential to understanding evolution. See Dietrich and Skipper (2012).

14. See Dietrich and Skipper (2012) and Skipper and Dietrich (2012). Pigliucci (2012) discusses useful distinctions between different kinds of landscapes.

15. See Simpson (1944).

16. See MacFadden (2005) as well as Simpson (1944) and Bell (2012).

17. Although scientists usually refer to them as Ammonoids, after the subclass Ammonoidea of cephalopods to which they belong, I will refer to them by their better-known common name of ammonites.

18. Nautilus was one but not the only discoverer of the principle of jet propulsion, which is also used by other water-living animals, such as jellyfish.

19. More precisely, the second quantity is the diameter of the umbilicus, the distance between the central axis and the inner wall of the outermost whorl, but the two are closely related. See Raup (1967) and Chapter 4 of McGhee (2007).

20. Center and right ammonite drawings after Figure 6 of Saunders et al. (2004).

21. See Chamberlain (1976, 1981).

22. See page 73 of McGhee (2007).

23. See page 360 of Chamberlain (1976), as well as Chamberlain (1981).

24. See page 75 of McGhee (2007), as well as Chamberlain (1981), McGhee (2007), and Saunders et al. (2004).

25. See Saunders et al. (2004) and McGhee (2007). Factors other than swimming efficiency, including buoyancy control and equilibrium control, may also matter, and the Ammonoid fitness landscape may thus be more complex than their analysis lets on. See McGhee (2007) and Chamberlain (1981).

26. See Hay-Roe and Nation (2007).

27. See Benson (1972).

28. See Brown (1981).

29. See Brown (1981), as well as Brower (1994, 2013).

30. See Haffer (1969), and also Knapp and Mallet (2003).

31. If we could "replay the tape" of life's evolutionary history, a thought experiment famously conducted by the late paleontologist Stephen Jay Gould, we would probably observe butterflies whose warning colors were just as effective as but completely different from those alive today. Their adaptive landscape would have different peaks.

32. See Majerus (1998), especially Chapter 6.

33. One of the reasons why the peppered moth captured the imagination of early population geneticists is that in its populations, genetic dominance—a phenomenon first described by Mendel—varies. In some populations, the light form is dominant over the black form—crosses of white and black moths yield mostly or only white moths—whereas in other populations the black form is dominant over the white form. The causes of this phenomenon of *dominance modification* led to a dispute between Fisher and Wright that lasted years. See Provine (1986).

34. Morgan's white allele is of special historical importance because it was the first allele with sex-linked inheritance—a different inheritance pattern in males and females—that pointed to the existence of sex chromosomes.

35. See Wright (1932).

Chapter 2: The Molecular Revolution

1. See Watson and Crick (1953).

2. More precisely, many hormones are short chains of amino acids also called peptides.

3. As I will discuss later in this chapter, through a process called alternative splicing cells can make more than one protein from a gene, such that the number of proteins in the human body is much greater than the number of genes.

4. And that is a gross underestimate of the size of the adaptive landscape on which *Drosophila* evolves, because the genomes of different fruit flies may differ in not just one but thousands or millions of different nucleotides. In addition, many genes are much longer than one thousand nucleotides, and many mutations can occur outside genes in the nonprotein coding regions of the genome.

5. See Mackenzie et al. (1999).

6. See Kauffman and Levin (1987).

7. 7 billion humans × 100 years × 356 days per year × 8.6×10^5 seconds per day equals 2.2×10^{20}. A landscape of four hundred gene loci with two alleles would have $2^{400} = 2.6 \times 10^{120}$ genotypes, or 1.2×10^{100} as many.

8. See equation (2) in Kauffman and Levin (1987) with $N = 15,000$.

9. See equation (4) in Kauffman and Levin (1987).

10. The number is given by the binary logarithm of 15,000; see page 23 of Kauffman and Levin (1987).

11. The notion of a protein space explored by evolution goes back at least to the biologist John Maynard-Smith. See Maynard-Smith (1970).

12. See Weinreich et al. (2006). The "conventional" beta-lactamase I am referring to is a member of a much larger family of proteins called TEM beta-lactamases, which share a similar amino acid sequence. It is the prototypical member of the family, also called the reference sequence. Four of the five mutations in question alter the amino acid sequence of the encoded protein, and the fifth is a regulatory mutation that occurs in the nonprotein coding part of the gene.

13. Even though only four amino acid changes are involved (and one DNA change in a regulatory region of the gene), the numbers still hold. TEM-1 is 263 amino acids long, and there are already 1.94×10^8 different ways of choosing just four amino acids for replacement. Each of them can be replaced by nineteen different other amino acids, so that for a mere four amino acid replacements, there are 2.5×10^{13} possible strings. That number has to be multiplied by another factor of three if one takes the fifth, regulatory change in DNA into account.

14. Under the condition of Weinreich's experiments, which used very strong selection for better variants, horizontal paths that do not affect fitness were also prohibited, even though they can play an important role in nature, as we shall see in Chapter 4.

15. Szendro et al. (2013) provide an overview of these experiments. I note that the genetic changes in such experiments need not all be single-letter changes in a protein. Some of them may affect the regulation of a gene rather than alter the letter sequence of an encoded amino acid string. Also, some changes are deletions or insertions of short pieces of a genome. However, the same principle applies here as to studies focusing exclusively on amino acid sequence variation. Different mutant alleles can arise in different orders, and only some of the resulting paths toward a fitness peak will be accessible.

16. Although it may not be the most important mechanism for such tuning. See Miranda-Rottmann et al. (2010). For this and other examples see also Graveley (2001).

17. Most splicing occurs in organisms like us—eukaryotes—and comparatively little occurs in bacteria.

18. See Hayden and Wagner (2012).

19. See Jimenez et al. (2013). This molecule is chemically very similar to the much more familiar ATP, containing a guanine in place of ATP's adenine.

20. See Badis et al. (2009), as well as Mukherjee et al. (2004) and Weirauch et al. (2014). The actual design of such a microarray is more complicated than described here, for example because some DNA words can contain gaps—regions that are not recognized by a regulator.

21. See Weirauch et al. (2014).

22. See Aguilar-Rodriguez et al. (2017).

Chapter 3: On the Importance of Going Through Hell

1. See Hawass et al. (2010).

2. See Alvarez et al. (2009).

3. Ibid.

4. I am referring to recessive diseases here, the most common diseases caused by mutations in a single gene, where both alleles of a gene in a diploid organism are required to have suffered a disease-causing mutation. Dominant diseases, where only one mutation suffices, would manifest themselves both in inbred and outbred individuals. It is also important to distinguish mutations in somatic tissues that do not contribute to reproduction from mutations in the germ line—the cells that get passed on to the next generation. Only the latter directly affect inheritance and future generations.

5. This also means that the best way of maintaining the lineage is to outbreed blue-eyed cats with non-blue-eyed cats, which would yield 50 percent offspring with blue eyes.

6. See Pusey and Packer (1987).

7. See Pusey and Wolf (1996).

8. For evidence supporting the Westermarck effect, see Shepher (1971), as well as Lieberman et al. (2003).

9. See Chapter 15 of Futuyma (2009), as well as Charlesworth and Willis (2009). It bears mentioning that some organisms, especially plants that have survived very small population bottlenecks through

the ability to self-fertilize, do not necessarily suffer from inbreeding depression, because many of their recessive deleterious alleles may have been purged already from their populations.

10. This is an important point. Inbreeding depression may be only one among many reasons why inbreeding is avoided in nature. In humans, for example, incest taboos can help forge family alliances. For relevant literature, see Charlesworth and Willis (2009), Pusey and Wolf (1996), and Szulkin et al. (2013).

11. The relevant mathematics comes from population genetics and, more specifically, from coalescent theory, which describes how many generations one must go back in time before a given number of individuals share a common ancestor. The details depend on whether one studies haploid or diploid organisms and whether one studies relatedness among all genes or just one gene in a genome, but the principle that this time depends linearly on the number of individuals in the population remains unchanged. See Hartl and Clark (2007).

12. See Coltman et al. (1999).

13. See Keller (1998) and Keller et al. (1994).

14. Wright also already proposed, in the form of his well-known shifting-balance theory, a process by which populations can descend from a peak through a valley to another (higher) peak of a complex fitness landscape. Some of the theory's details are complex and controversial, but one central element is simple and widely accepted: genetic drift can help populations traverse adaptive valleys. See Chapter 9 of Provine (1986).

15. I simplify the genetics and evolution of eye color here for the sake of having a concrete example to illustrate the concept of genetic drift. Although high school biology classes sometimes discuss eye color as an example of a trait influenced by a single gene, it is actually affected by multiple genes, some of them more important than others. An especially important gene is OCA2, which is involved in the synthesis of brown pigment in the iris, and a single mutation that changes this gene's expression is sufficient to turn brown eyes into blue. See Eiberg et al. (2008). It is not clear that eye color is a neutral trait— i.e., not affected by natural selection—because blue eyes have spread rapidly since their origin some ten thousand years ago. Also, eye color affects the incidence of diseases such as macular degeneration and

uveal melanoma. See Sun et al. (2014). If I nonetheless use eye-color alleles to explain the phenomenon of genetic drift (rather than more unambiguously neutral alleles) it is because few neutral alleles affect a human trait as plainly visible as eye color.

16. The reason is that a random letter change in a long DNA text is much more likely to create a new variant than to revert a mutation that has occurred previously.

17. This holds as long as selection does not affect the sampling. For example, a specific class of "selfish" genes can promote their own propagation to the next generation at the expense of the organism's genetic health in a phenomenon called meiotic drive. See page 290 of Futuyma (2009).

18. Even though eye color is technically a polygenic trait—influenced by multiple genes—brown eyes are usually dominant over blue eyes, meaning that only one of the alleles in an individual's genome would have to be "brown" for the iris to be brown, whereas both alleles would have to be "blue" for blue eyes. Even so, because 50 percent of populations would get fixed for the blue allele, in those populations all individuals would have blue eyes.

19. I am using here two basic insights from population genetics. The first is that a gene evolving only under the influence of genetic drift, with a frequency of p in one generation, will have a random allele frequency in the next generation whose mean is p and whose variance is of the order of $p(1-p)/N$, where N is the population size. If one were to choose the standard deviation as a measure of dispersion, allele frequencies would fluctuate by an amount that is inversely proportional not to N but to the square root of N. The second insight comes from coalescent theory, which shows that the amount of time one has to go back in time to find the common ancestor of two (or all) alleles in the population is of the order N. I note that all these expressions are for haploid organisms. For diploids, N needs to be replaced by $2N$, which does not, however, affect the order-of-magnitude argument of the main text. See Hartl and Clark (2007).

20. Alleles causing dominant disease, where only one of two copies needs to be mutated for the disease to occur, cannot spread as easily through genetic drift, because natural selection would prevent their spreading. The reason recessive alleles can spread through drift is that

they only manifest their negative effects when in two copies, which is very unlikely to happen unless they already have a large frequency. More generally, most genetic diseases are complex diseases caused by mutations in multiple genes, where any one mutation may contribute very little to disease risk and can thus spread far through a population by drift alone. Also, many naturally occurring mutations have fitness effects that are deleterious, but very weakly so, such that natural selection may delay but does not prevent their spreading by genetic drift. See Hartl and Clark (2007).

21. More precisely, they do not shuffle genomes every generation like we do. However, they engage in other forms of sexual reproduction that affect only part of their genomes and that do not occur every generation, such as bacterial conjugation. See Griffiths et al. (2004).

22. In the interest of simplicity, I have taken several liberties with the landscape concept here. First, as opposed to Chapter 1, where the units of study—individuals on the landscape—are for the most part genotypes, here the unit of study is an entire population. Viewed as an object on the landscape—one can think of this object as also representing the center of mass of a group of individuals—natural selection drives this object uphill. But as a result of genetic drift, the location of this object fluctuates because the population's allele frequencies fluctuate. So, strictly speaking, it is not the landscape itself but the object that is being shaken. A better but more technical analogy is that of a particle under Brownian motion with a diffusion coefficient proportional to the inverse of population size.

23. Technically speaking, I am referring to a general observation from population genetics that the selection coefficient of an allele must be smaller than approximately the inverse of the population size (which is proportional to the generation-to-generation variation in neutral allele frequency) in order for drift to be able to overcome the pull of selection. See Hartl and Clark (2007). The exact number depends on whether organisms are haploid or diploid (where population size N has to be replaced by $2N$), on which aspect of fitness one considers, and on the units in which this aspect of fitness is measured. Also, I note, for the cognoscenti, that whenever I am discussing population size, I am considering what population geneticists call the effective size of a population, which is the relevant quantity for

genetic drift, but may be smaller than a population's census size. See also Lynch (2007).

24. See Eyre-Walker and Keightley (2007) and Chapter 5 of Freeman and Herron (2007).

25. That's because the (effective) population sizes of bacteria usually exceed 10^8 individuals. See, for example, Chapter 4 of Lynch (2007).

26. See Sun et al. (2014).

27. See Table 1 on page 219 of Whittaker and Fernandez-Palacios (2007).

28. See Sulloway (1982).

29. See pages 228–229 and Table 9.3 of Whittaker and Fernandez-Palacios (2007).

30. The oldest islands on Hawaii and Galápagos are Kauai and Española. See Geist et al. (2014), as well as page 220 of Whittaker and Fernandez-Palacios (2007), for relevant dating information. It is important to be aware that in volcanic archipelagos, islands can arise and become submerged again, such that an archipelago may be older than its oldest island visible today. Molecular clock dating can be used to estimate the age of the oldest organismal lineage on an archipelago, which shows, for example, that on Hawaii few lineages are older than ten million years, a short amount of evolutionary time.

31. The examples in this paragraph and others can be found in Chapter 9 of Whittaker and Fernandez-Palacios (2007).

32. Island life in particular shows recurrent patterns of change that include gigantism in plants and some animals, as well as reduced dispersal; for example, through flightlessness. See Grant (1998).

33. See Montgomery (1983).

34. Most scientists believe that population bottlenecks on islands play an important role in adaptive radiations. See Chapter 7 of Whittaker and Fernandez-Palacios (2007). However, they are not the only factor. Just as important is the reduction of competition that is experienced by species that occupy previously empty ecological niches when they colonize an island. More intense competition often means more stringent selection, and reduced competition on islands means that the intensity of selection is reduced, another case in point that curtailing selection—by whatever means—can facilitate innovation.

35. See page 453 of Lynch (2006), as well as Carbone and Gittleman (2002).

36. Again, when I refer to population size here and throughout, I am referring to what population geneticists call the effective population size, which is typically much smaller than the census population size and reflects the fraction of individuals or genes that contribute to the next generation's gene pool. It is influenced by several factors, including the mode of reproduction and variation in census population size over time. See Hartl and Clark (2007).

37. An additional factor whose effect is difficult to predict is that the kind of selective pressures we are subject to are changing. On the one hand, we suffer to a greater extent from diseases related to our modern lifestyle, such as type II diabetes. On the other hand, medicine has made huge progress in helping us compensate for some defects caused by our genetic heritage.

38. In another difference from eukaryotic organisms like us, several of *E.coli*'s regulatory proteins are themselves part of the RNA polymerase that transcribes genes. In the interest of brevity, I am omitting a few other roles of non–protein coding DNA, such as antiviral defense or assistance in the initiation of DNA replication, because gene regulation is so prominent among them.

39. To be precise, the average amount of non–protein coding DNA found between two human genes exceeds one hundred thousand base pairs (bps) and is thus almost one thousand times greater than that found between two *E.coli* genes. The calculation is based on 2.9×10^9 base pairs as the size of the human genome, and on twenty-four thousand genes with an average protein-coding length of 1330 base pairs. See Table 3.2 of Lynch (2007). Some human genes are millions of base pairs apart.

40. A substantial fraction of vertebrate non-coding DNA is transcribed into RNA that is not translated into protein, some of which may also help regulate genes.

41. See Lynch and Conery (2000).

42. Other costs include that of transcribing the new gene into RNA, as well as the—comparatively small—cost of manufacturing the DNA building blocks of the additional DNA. The cost is not the same for all

genes and depends on the amount of RNA and protein manufactured. See Wagner (2005, 2007).

43. Another difference is that in higher organisms, the energy costs of gene expression may be less important in determining reproductive success than many other factors, such as mobility, cognitive abilities, and attractiveness to mates.

44. See pages 60–61 of Lynch (2007). Mutations that inactivate gene duplicates are not the only—and not even the most important—source of pseudogenes. Another is a mechanism called retroposition that creates gene duplications and is very different from the DNA recombination and repair I discuss in the main text. Retroposition is a process in which the RNA transcribed from a gene is transcribed back into DNA by an enzyme called reverse transcriptase. Often, the resulting DNA is not a complete copy of the gene, or it integrates into a location of a genome that does not contain the necessary regulatory DNA words needed to transcribe the gene. Such genes are effectively dead on arrival and form a large reservoir of so-called retropseudogenes in our genome.

45. See Dawkins (1976).

46. There are multiple different kinds of mobile DNA, also referred to as transposable elements. They include transposons, long terminal repeat elements, as well as long and short interspersed nuclear elements. See pages 56–60 of Lynch (2007).

47. This selection can take different forms and has led to the "domestication" of some elements, favoring organisms whose mobile DNA causes mutations that are relatively harmless, for example, by inserting into gene-poor regions where its insertions are not likely to be damaging.

48. See Chapter 7, page 168, of Lynch (2007).

49. See pages 174–179 of Lynch (2007).

50. See page 178 of Lynch (2007) and Figure 5 of Lynch (2006).

51. See page 57 of Lynch (2007).

52. See pages 56–60 of Lynch (2007). Short interspersed nuclear elements are particularly abundant in our genome, which contains more than 1.5 million of them. They can only transpose passively, using the transposition enzymes from other mobile DNA to do so. They are the ultimate DNA parasites.

53. See Lynch (2007).

54. See Gilbert (1978).

55. See Lynch and Conery (2003) and pages 256–261 in Lynch (2007). My discussion of introns is based on so-called spliceosomal introns, which are characteristic of eukaryotes and do not exist in prokaryotes—another aspect of their simpler genome organization. And when I refer to microbes here I mean eukaryotic microbes, for example, unicellular fungi like baker's yeast.

56. See Table 3.2 of Lynch (2007).

57. See page 51 and Table 3.2 of Lynch (2007). The simpler genome organization of bacteria also has advantages. It allows shorter generation times and facilitates the ability to transfer genes horizontally.

58. See Tables 3.1 and 3.2 of Lynch (2007).

59. See Chimpanzee Sequencing and Analysis Consortium (2005).

60. None of this implies that natural selection is unimportant. Selection is essential to reliably ascending fitness peaks in adaptive landscapes.

Chapter 4: Teleportation in Genetic Landscapes

1. See Hardison (1999), as well as Aronson et al. (1994). They are different solutions in the sense that their amino acid strings are very different, but they bind oxygen by the same mechanisms. Essentially, they are different texts expressing the same "meaning."

2. See Wagner (2014).

3. See Hayden et al. (2011).

4. See Figure 6 of Bershtein et al. (2008). See also Wu et al. (2016) for a different kind of experimental demonstration showing that adding dimensions to an evolutionary journey may facilitate adaptation.

5. Similar networks of high-elevation ridges also occur in mathematical models of speciation—that is, of the evolution of reproductive isolation—where mathematical biologist Sergei Gavrilets calls them "holey" adaptive landscapes. See Gavrilets (1997).

6. The phrase is so well-known that it has its own Wikipedia page (https://en.wikipedia.org/wiki/Beam_me_up,_Scotty), even though Kirk, played by William Shatner, never used this exact phrase.

7. The assumption is that the two chromosomes in a pair come from unrelated individuals. If the mother and father of the individual

whose homologous chromosomes are compared are related (i.e., they come from an inbred family), the number of differences could be much smaller. Also, the incidence of pairwise differences varies widely across different regions of the human genome. See Jorde and Wooding (2004).

8. To be precise, three billion is the approximate number of nucleotides in one set of twenty-three chromosomes, and to get the number of nucleotides in all chromosome pairs, this number would need to be doubled.

9. I am neglecting here that a small number of approximately 30–40 (germ-line) mutations occur during the life cycle of each human, as this number of differences pales in comparison to the amount of genetic change introduced by recombination between a typical mother and father. See Campbell and Eichler (2013).

10. The calculation is based on 1.5×10^6 steps and 2.5 feet per step, which amounts to 710 miles. If the distance from one "end" of the landscape to the other were given by the maximally possible number of nucleotides that could differ between two genomes (which is approximately equal to 3×10^9 for a genome of the same size of the human genome), then that distance would translate to 3×10^9 steps, which equals 7.5×10^9 feet or 1.42×10^6 miles (2.28×10^6 kilometers). Comparing that distance of more than two million kilometers to the distance between the earth and the moon (at a mere 384,400 kilometers) is another way to appreciate how vast this landscape is.

11. Such instant speciation through hybridization can sometimes take place when plants double their chromosome numbers, which can occur spontaneously as a result of DNA-replication or cell-division errors. See Futuyma (2009). Many hybrids may be less well adapted to any habitat than their parents are, but a few can discover entirely new "lifestyles." See Arnold et al. (1999) and Arnold and Hodges (1995).

12. See Pennisi (2016).

13. See pages 492–493 of Futuyma (2009), as well as Rieseberg et al. (2007).

14. See Lamichhaney et al. (2018) and Lamichhaney et al. (2015), as well as Grant and Grant (2009) and Pennisi (2016). The descendants of the large bird are also an example of successful inbreeding because only two siblings survived the drought, mated, and founded

a family lineage whose members largely reproduce with each other. Darwin's finches are only a few among a growing number of species where it has become clear that hybridization is rampant. Its abundance softens "hard" boundaries between species because even species that were previously thought to be reproductively isolated—the very definition of a biological species—often turn out to exchange material in a process called introgressive hybridization, which need not lead to speciation. It takes place when hybrid organisms reproduce over multiple generations with members of either parental population. Allele combinations that help the hybrid survive in a new habitat can then become preserved in its genome. This phenomenon has been documented in diverse organisms, including some fungi that are involved in cheese making, malaria mosquitoes, and American gray wolves. See Arnold and Kunte (2017), as well as Pennisi (2016), Ropars et al. (2015), Norris et al. (2015), and Anderson et al. (2009).

15. See Bushman (2002). The process usually terminates before all genes in a genome have been transferred. It tends to transfer genes jointly that are in close proximity on the bacterial DNA, a fact that has been used to map genes in the *E.coli* genome. Because the transferred genes often include those encoding the ability to make a sex pilus and donate DNA, the genome transfer machinery may be spreading "selfishly" among bacteria. But in doing so, it also transfers other genes from the donor that may prove useful to the recipient. I also note that not all horizontal gene transfer involves bacterial sex (conjugation). Other recombination mechanisms include the uptake of naked DNA (transformation) and the shuttling of DNA from one cell to another through infectious viruses (transduction). Horizontal gene transfer also occurs between bacteria and plants, fungi, or animals, as well as among the latter three classes of organisms, although the mechanisms are not always understood. See Arnold and Kunte (2017).

16. Recombination can occur between bacteria that differ in 10 percent or more of their DNA text, compared to the average of 0.1 percent in humans. See Fraser et al. (2007). By comparison, for example, genomic diversity in sunflowers is greater than that in humans but is also below 1 percent of DNA text divergence. See Pegadaraju et al. (2013). Bacteria have different generation times, smaller mutation rates, and genomes more dense with genes. See Lynch (2007). For

these reasons, the same degree of genomic difference between two bacteria and two higher organisms does not necessarily translate into the same amount of time since their most recent common ancestor lived.

17. See Gelvin (2003), as well as Robinson et al. (2013).

18. Some simple animals are in fact capable of photosynthesis, but they engage in symbioses with other organisms that provide this ability for them, and they do not use the synthesized carbohydrates as their only source of nutrition. An intriguing question is why more of them do not engage in these symbioses. See Smith (1991).

19. See Copley et al. (2012), Russell et al. (2011), and Maeda et al. (2003), as well as Hiraishi (2008).

20. I note that many antibiotic-resistance traits are only encoded by a single gene, which makes their spreading especially easy.

21. Some viruses, like the human immunodeficiency virus (HIV), undergo recombination while they evolve inside patients, but it would be hard to match the density and diversity of recombining molecules in a test tube.

22. More precisely, they (and DNA shuffling) use a heat-stable polymerase that is needed for the polymerase chain reaction. This reaction is an important tool of molecular biology to make many copies of a DNA sequence of interest. See Stemmer (1994).

23. See Crameri et al. (1998). They improved an enzyme that can cleave moxalactame, another cephalosporin antibiotic like cefotaxime.

24. See Ness et al. (1999) and Raillard et al. (2001), as well as Crameri et al. (1997).

25. More precisely, the organisms I am referring to do not reproduce sexually.

26. See Judson and Normark (1996).

27. See Flot et al. (2013).

28. However, in molecules like proteins one can create many recombinants experimentally and evaluate what percentage of them are functional proteins. See Drummond et al. (2005).

29. See Drummond et al. (2005) and Martin and Wagner (2009), as well as Hosseini et al. (2016). While this work simulates the effects of recombination on DNA, a limited amount of experimental work on proteins in Drummond et al. (2005) reaches the same conclusion.

Chapter 5: Of Diamonds and Snowflakes

1. See pages 81–114 of Gerst (2013).

2. See Kroto (1988) and Kroto et al. (1987).

3. See Smalley (1992).

4. Their original discovery is described in Kroto et al. (1985). For early reviews see Smalley (1992) and Kroto (1988). Bucky-balls are one of multiple allotropes—different physical forms—of carbon that also include graphite and diamond.

5. See Cami et al. (2010), as well as Berne and Tielens (2012) and Garcia-Hernandez et al. (2010).

6. See Campbell et al. (2015).

7. To be precise, van der Waals first postulated the existence of such a force.

8. More specifically, four atoms form a tetrahedron, and five atoms form a triangular bipyramid, which can be obtained by joining two tetrahedra. The numbers I cite in this paragraph come from experimental observations and theoretical calculations by Meng et al. (2010). I note that the number of valleys and how it scales with the number of atoms depends on the kinds of atoms and forces considered. See Wales (2003) and Berry (1993). Also, the numbers reported here come from free-energy landscape calculations, which take not only potential energy but also entropy into account. Entropy refers to the number of configurations that a given number of atoms can assume, and the laws of thermodynamics tell us that atoms, left to their own devices, will tend to maximize the number of configurations they can assume. In other words, while minimizing their potential energy, atoms will also tend to maximize their entropy. In combination, the two principles lead to even more complex landscapes. On a different note, it is worth pointing out that the most stable arrangements are not highly regular in all materials. They can even be irregular, for example in gold. See Michaelian et al. (1999). High regularity and symmetry mean that multiple minima correspond to the same stable atomic configuration but with permuted identities ("labels") of atoms. All these factors—strength of force, entropy, symmetry—influence landscape structure, but they do not affect the central principle that the complexity of potential and free-energy landscapes increases exponentially with the number of atoms.

9. For example, in a cluster of merely thirty-two potassium chloride molecules, there are at least ten billion more minima of high potential energy corresponding to amorphous structures than there are low-lying minima corresponding to stable structures with an atomic configuration similar to the cubic arrangement familiar from rock salt. See page 2389 of Berry (1993).

10. See Oliver-Meseguer et al. (2012), Corma et al. (2013), as well as Michaelian et al. (1999). I note that the distinction between a molecule and a cluster is not clear-cut. See, for example, Chapter 1.2 of Wales (2003). The sizes of the gold clusters I mention are in the hundreds of picometers (10^{-12} meters) rather than nanometers (10^{-9} meters), such that one could assign them to the realm of "pico-technology." See Cartwright (2012).

11. See Wales (2003).

12. Here it is worth pointing out some simplifications in the main text. For example, the importance of cooling in crystallization comes not only from the efficient exploration of an energy landscape, but also from the fact that the solubility of many solutes decreases with decreasing temperature, which makes more and more solute molecules available for crystal growth. This is also why (slow) evaporation of the solvent, which reduces its volume and thus increases the solute's concentration, is a commonly used strategy for crystallization. If cooling or evaporation occurs too fast, many solute molecules will precipitate in amorphous clumps. Also, when a crystal forms, not all its constituent atoms or molecules simultaneously explore the crystal's energy landscape. Instead, a crystal begins to form through nucleation, a process in which some of a solute's molecules associate in the correct (minimum energy) configuration either by themselves or as prompted through an impurity, such as a dust particle in the solution. The crystal then grows from such a nucleus as more and more molecules aggregate. This means that the energy landscape is not explored haphazardly but rather in preferred directions that are given by how growth occurs. This is a specific example of a more general phenomenon in self-assembling molecular structures: kinetics can facilitate or hinder their formation. But even when our "marble" explores an energy landscape along some preferred direction, it may encounter

shallow valleys separated by peaks—suboptimal and imperfect molecular arrangements—and so heat-induced jitters are still important.

13. Those where at least some atoms are covalently bound are known as covalent crystals. A prominent example is diamond, where each carbon atom is bound to four adjacent ones in a highly regular arrangement that yields the octahedral symmetry of diamond.

14. See Smalley (1992) for a discussion of the temperatures at which bucky-balls form and for the importance of (relatively) slow cooling for a high yield of bucky-balls.

15. See page 501 of Wales (2003) for the short assembly time of bucky-balls. It also testifies to the importance of assembly kinetics, the exploration of an energy landscape along preferred directions, which comes from the observation that bucky-balls do not assemble from scratch in one step but instead build themselves piece by piece from smaller yet already highly regular molecules. See Kroto (1988) and Smalley (1992).

16. An additional complication is that some materials are polymorphic; that is, they can form alternative but similarly stable crystal structures.

17. This is only one of the many complexities involved in growing snowflakes. See Libbrecht (2005). It is another example where self-assembly kinetics is important.

18. Some defects of bucky-balls are visualized in Chapter 8.6 of Wales (2003).

19. This statement about the majority of carbon atoms refers to the fraction of carbon atoms that get bound in large clusters, more than 50 percent of which can be bucky-balls under the right conditions. See Kroto et al. (1985). Even overall yields exceeding 20 percent, as have been reported, are remarkable. See page 108 of Smalley (1992).

Chapter 6: Creative Machines

1. See Biery (2014).

2. For the vehicle routing problem with n customers, the depot is not included in the city count, so the basic count of tours is $n!=1\times2\times..\times n$. One cannot generally assume that a route and the route obtained from

it by reversing the customer order are equally long. For example, the routes from the depot to the first and last customer may have different lengths, or one-way streets may exist. This means that the number $n!$ of routes cannot be reduced further by taking such "symmetries" into account. For the closely related traveling salesman problem, the number of routes is $(n\text{-}1)!$, because one of the n cities (e.g., the starting point) is arbitrarily chosen (it corresponds to the "depot"), and the remaining $(n\text{-}1)$ cities can be permuted at will. For both problems, the number of solutions grows faster than exponentially.

3. See Chapter 2 of Cook (2012) for an overview of the problem's history. The term *traveling salesman problem* was not used until the middle of the twentieth century. The problem itself is often associated with the Irish mathematician Sir William Rowan Hamilton, who studied a special mathematical instance of the problem, namely how to visit all twenty vertices of a dodecahedron in a continuous path along the dodecahedron's edges. His name is attached to the notion of a Hamiltonian circuit, the problem of finding a circuit through a network (a graph in mathematical language) that visits each of the network's nodes exactly once. Solving the TSP is tantamount to finding a shortest Hamiltonian circuit through such a graph.

The many solutions for problems like these are not the only reason they are hard. There are other problems that have many solutions and that are easy to solve, because the best solution is easy to find among all solutions. The reason is that the landscape describing these solutions is simple. Among such easy problems is the minimum spanning tree problem. See Chapter 3 of Moore and Mertens (2011). More generally, in computer science, hard problems are also referred to as NP ("non-deterministic polynomial-time")-hard, a term that refers to how the amount of time needed to solve a problem scales with problem size and distinguishes these problems from easier ones that are solvable in an amount of time that scales polynomially with problem size. Harder problems are those with more rugged solution landscapes. I note that the question of how to distinguish rigorously between easy and hard problems is itself one of the deepest open problems in computer science and mathematics. See, for example, Chapter 6 of Moore and Mertens (2011).

4. See Matai et al. (2010), Chapter 3 of Cook (2012), as well as Chapter 4 of Rhodes (1999).

5. For an algorithm specifically aimed at minimizing the carbon footprint of a vehicle route, see Liu et al. (2014).

6. This is only one among multiple kinds of moves that are possible in the solution landscape. Another one, also referred to as 2-opt, cuts two parts of a route that cross each other and then swaps these parts between the two customers they serve. 2-opt helps avoid self-crossing routes, which are inefficient. See Section 9.10 of Moore and Mertens (2011).

7. More precisely, a greedy algorithm is one that always chooses the best among a (limited) number of options. The two options here are the original and the swapped customer order, and the algorithm chooses the one that creates the better solution. In the context of landscape search, algorithms are often referred to as greedy if they ascend the steepest slope to a peak, but this definition presupposes that the algorithm can evaluate all possible next steps, which can require a lot of computation if there are many such steps. I also note that there can be multiple greedy algorithms for any one problem. A better-known greedy algorithm for the vehicle routing problem first visits the customer closest to the depot, travels from there to the closest customer, and so on. It needs to evaluate the distance to all possible customers. In the traveling salesman problem, a specific algorithm that grows many sub-paths simultaneously is known as *the* greedy algorithm, even though it is only one among many such algorithms. See page 67 of Cook (2012).

8. Combinatorial optimization problems have been known much longer than just for half a century. The TSP, for example, originated in the nineteenth century. However, even moderately large TSPs could not be solved efficiently before the advent of computers and efficient algorithms. My reference point in time here is one of the earliest computational milestones, the 1947 discovery of a technique called the simplex algorithm, which is able to solve many combinatorial optimization problems that are not truly hard. See Dantzig (1963).

9. See Sibani et al. (1993), as well as Figure 4 and Table 5 of Hernando et al. (2013), which show these numbers for the traveling

salesman problem. (The corresponding routing problem of the same size would have even more local minima.)

10. See Chapter 10 of Glover and Kochenberger (2003).

11. One can take two different perspectives on the nature of the "marble" in evolution, real or simulated. A marble can represent an individual, as in the image I use in the main text. It can also represent an average location or center of mass for an entire population. Strong genetic drift is analogous to high temperature, in that it causes fluctuations in the population's center of mass.

12. See Turing (2013).

13. See Holland (1975) and Mitchell (1998). Holland was not the first to work with such algorithms. One important predecessor was the German scientist Ingo Rechenberg, whose work is less widely known. See Rechenberg (1973). Various flavors of such algorithms have been developed, and they come under a variety of names, such as genetic programming and evolutionary algorithms. See, for example, Koza (1992). For simplicity, I will here restrict myself to the term *genetic algorithms*.

14. For example, two genetic algorithms for the vehicle routing problem that use fifty or fewer individuals are described by Prins (2004), as well as by Baker and Ayechew (2003).

15. To be sure, this kind of recombination may not work exactly as in biology. For example, recombining two DNA chromosomes is like swapping the customers in the first half of one delivery route with the customers in the second half of another route. The problem is that the resulting child might not even be a valid route: it might not visit all customers, or it might visit some customers twice. The best way to simulate recombination depends on the problem and its encoding in an algorithm's genome.

16. See Koza et al. (2003). You may have noticed that I did not mention the analogue of a third feature that helps biological evolution explore an adaptive landscape—the sprawling network of ridges of approximately equal altitude from Chapter 4. That's because such networks remain poorly explored in the solution landscapes of computer science, but we have hints that they could help solve difficult problems there as well. See Raman and Wagner (2011), as well as Banzhaf and Leier (2006).

17. See Glover and Kochenberger (2003), as well as Moore and Mertens (2011). Large instances of problems like the VRP or TSP are often solved by a combination of algorithms. One might start with a greedy algorithm, then change the resulting route through the strategic swapping of edges to avoid self-crossing paths, perturb the structure of the problem itself to solve a similar but easier problem in order to obtain tight upper and lower bounds on the solution, then try to map the solution of the easier problem onto a valid solution for the harder one, which can then be improved further, and so on. See Chapter 9 of Moore and Mertens (2011). These multistep procedures are often not conceived with the structure of a solution landscape in mind, but their steps help avoid getting trapped in its local minima. Such procedures also often take advantage of mathematical insights into the structure of a problem, whereas general-purpose algorithms like genetic algorithms can be applied to any problem. That's both an advantage and a limitation, as general-purpose algorithms are usually not as efficient as custom-made algorithms for very specific problems.

18. The tours I discuss refer to the traveling salesman problem, the close relative of vehicle routing with one vehicle, one depot, and no further constraints. The 666-city tour was reported in Holland (1987). For a review of this and other impressive computational records see Chapter 8 of Cook (2012).

19. Once these building blocks are chosen, the engineer designing a genetic algorithm also faces another important and difficult choice, namely how to evaluate the performance of a solution by choosing the right "fitness function." This choice may be simple for problems like the vehicle routing problem, but it can be difficult for other problems, especially if they involve complex technologies like engines or airplanes whose performance has multiple aspects.

20. See Koza et al. (1999).

21. See Keane et al. (2005). See also Keats (2006) and Koza et al. (2003). For the effects that creative machines may have on patent and intellectual property law, see Plotkin (2009).

22. See Wang et al. (2013).

23. See Keats (2006).

24. See Hornby et al. (2011).

25. See Schmidt and Lipson (2009). The authors used an evolution-ary computation method called symbolic regression, which combines mathematical building blocks into equations that aim to describe ex-perimental data.

26. See page 1 of Plotkin (2009).

27. Not all of them are unique, of course, but the same holds for biological evolution, which has come up with similar solutions to some problems multiple times through what biologists call convergent evolution.

28. See Van Tonder et al. (2002), as well as Taylor et al. (2011) and Pachet (2012).

29. See Fernandez and Vico (2013).

30. See Muscutt (2007), as well as Cope (1991) and Adams (2010).

31. See Adams (2010).

32. See Muscutt (2007).

33. See Johnson (1997).

34. See Muscutt (2007).

35. See Cope (1991) and Adams (2010).

36. See Fernandez and Vico (2013), as well as http://www.geb.uma.es /melomics/melomics.html. For some press coverage on the computer and algorithm behind this composition see Smith (2013) and Ball (2012).

37. See Pachet (2008), as well as https://www.francoispachet.fr /continuator/. I note that the task of improvising in another person's style is less difficult than creating compositions from scratch. See, for example, Fernandez and Vico (2013).

38. See Levy (2012).

39. See Podolny (2015).

40. See Levy (2012).

41. See Clerwall (2014).

42. See Constine (2015).

Chapter 7: Darwin in the Mind

1. For a detailed account of the painting's creation, together with the historical context, see Chipp (1988).

2. See Chipp (1988), but also Weisberg and Hass (2007), as well as Weisberg (2004) and Simonton (2007a).

3. See Simonton (1999) and Simonton (2007a).

4. The ideas of Bain and other early thinkers on Darwinian creativity are summarized in Campbell (1960). Darwin's great contribution lay in recognizing the importance of natural selection and common descent, but not in elucidating the origin of new variation by random mutation. He did not know where new variation came from and freely admitted his ignorance.

5. See James (1880).

6. See Campbell (1960). Campbell's term referred to the inner workings of the human mind, but it applies in equal measure to the growth of human knowledge. The philosopher Karl Popper recognized this when he said that "the growth of our knowledge is the result of a process closely resembling what Darwin called natural selection." See page 26 of Simonton (1999).

7. See page 190 of Dehaene (2014).

8. Simonton's analysis of Picasso's *Guernica* was not uncontroversial. Much of the controversy revolved around the question of how unrelated the imagery and sketches were to anything that had come before them, and thus how free Picasso's associative process really was. See Weisberg (2004) and Dasgupta (2004), as well as Simonton (2007b) and Weisberg and Hass (2007). But as I discuss in the main text, even blind variation in biological evolution builds on previous variation and is thus not free in this sense. See Wagner (2012). I note that an analogous tension exists in evolutionary biology, where students of organismal development point out that development constrains the kind of variation that DNA mutations can generate. See Maynard-Smith et al. (1985) for the concept of constrained evolution.

9. See Weisberg and Hass (2007).

10. See pages 331 and 340 of Simonton (2007a).

11. See John-Steiner (1997).

12. These and other quotes can be found on pages 26–34 of Simonton (1999).

13. As quoted in Lohr (2007).

14. See Plunkett (1986) and Rosen (2010). For further examples see Wagner (2014) and pages 35–39 of Simonton (1999).

15. See page 84 of Simonton (1988). Needless to say, using citation counts as the only indicator of influence can be highly misleading.

There are huge differences among fields in the average number of citations a work receives. For example, research in biology gets more citations than research in mathematics. Citations to new research methods tend to be more numerous than to other kinds of research. And some unfortunate authors receive many negative citations, references to how research should not be done. Other issues with the use of citation counts are summarized on page 85 of Simonton (1988).

16. See page 84 of Simonton (1988).

17. See Lariviere et al. (2009).

18. See Simonton (1985). It is worth noting that Koehler, one of the fathers of Gestalt psychology, was opposed to the idea that trial-and-error could explain his chimpanzees' problem solving. See page 389 of Campbell (1960) for a discussion of his objections and those of others against Darwinian creativity. The key sticking point seems to be again the question of whether candidate solutions are created from scratch or build on what is already there, for example, pre-existing concepts that require some previous insight into the problem to be solved.

19. See Simonton (1977).

20. Page 92 of Simonton (1988) discusses this and other examples.

21. See page 93 of Simonton (1988).

22. See pages 154–155 of Simonton (1999).

23. Cited on page 186 of Simonton (1994).

24. See Stern (1978).

25. See pages 184–187 of Simonton (1994), as well as Stern (1978) and Sinatra et al. (2016). The latter study also shows that different scientists differ in their potential to create highly important work in their lifetime, even though the impact of any one work is hit or miss.

26. I am referring to one of the earliest successful theories of color vision known as the Young-Helmholtz theory, which proposed the trichromatic nature of our vision.

27. See Chapter 3 of Palmer (1999) and Chapter 1 of Gärdenfors (2000).

28. See Chapter 31 of Kandel et al. (2013) and Chapter 2 of Gärdenfors (2000).

29. See Gärdenfors (2000).

30. See Simonton (2007a).

31. See page 356 of Weisberg and Hass (2007).

32. See pages 45–46 of Padel (2008).

33. See page 14 of Hadamard (1945).

34. See Campbell (1960).

35. See Chapter 1 of Wales (2003).

36. See page 282 of von Helmholtz (1908).

Chapter 8: Not All Those Who Wander Are Lost

1. See page 16 of Bateson and Martin (2013).

2. Ibid., 17.

3. See Caro (1995) and Henig (2008).

4. See references on page 342 of Caro (1995).

5. See Harcourt (1991).

6. See Cameron et al. (2008).

7. See Fagen and Fagen (2009).

8. Practicing this behavior also has other benefits. For example, females who participate in non-conceptive sexual behavior also invest more into the first eggs they lay. See Pruitt and Riechert (2011).

9. See Spinka et al. (2001) and Henig (2008). Both sources also discuss a number of alternative hypotheses about the purpose of animal play, which comprises very heterogeneous activities.

10. See Wenner (2009).

11. See page 31 of Bateson and Martin (2013).

12. Cited on page 247 of Root-Bernstein and Root-Bernstein (1999).

13. See pages 58–61 of Bateson and Martin (2013).

14. See page 82 of Jung (1971).

15. See page 198 of Martin (2002).

16. For these and other examples, see pages 198–204 of Martin (2002).

17. Digital assistants can be used to the same effect, as can more sophisticated experiments that test reaction times. See Jackson and Balota (2012).

18. See Kane et al. (2007), as well as Jackson and Balota (2012), Killingsworth and Gilbert (2010), and Christoff (2012).

19. See Mooneyham and Schooler (2013).

20. See pages 13–14 of Hadamard (1945).

21. See Baird et al. (2012).

22. See Mrazek et al. (2013).

23. See Schooler et al. (2014) and references therein.

24. See Schooler et al. (2014). One might think that the two opposing mental processes are instead analogous to selection and mutation. However, it appears that our minds incessantly create spontaneous associations, and what mind-wandering does is to permit remote rather than close associations to come to the fore. See Baror and Bar (2016). The importance of a balance between opposing forces becomes especially clear when it is seriously out of whack in mental illnesses like schizophrenia and psychosis, whose symptoms include thought disorders such as a rambling form of speech known as word salad. Low latent inhibition and low negative priming, phenomena associated with creative personality traits in some experiments, have also been associated with schizophrenia. See Lubow and Gewirtz (1995), Beech and Claridge (1987), as well as Lubow et al. (1992). More generally, creativity has been associated with psychoticism, a dispositional trait causing susceptibility to psychotic symptoms. See Eysenck (1993). Perhaps more frequent, however, are mood disorders such as depression. See Chapter 4 of Runco (2014). Common wisdom had it for centuries that outstanding creativity effectively comes with madness. "Great wits are sure to madness near allied, and thin partitions do their bounds divide" is how the seventeenth-century poet John Dryden expressed it. Alas, that wisdom is a bit dated. In the words of psychologist Mihaly Csikszentmihalyi, who has interviewed countless eminent creators, "the reigning stereotype of the tortured genius is to a large extent myth." See page 19 of Csikszentmihalyi (1996). See also Simonton (2014). Even a mind with outsized creative powers can be mentally healthy and happy.

25. See page 2 of Bateson and Martin (2013).

26. As quoted in the TED talk "Tales of Creativity and Play" by Tim Brown, CEO of the design company IDEO, available at http://www .ted.com/talks/tim_brown_on_creativity_and_play.

27. Successful creative teams share a property called psychological safety, which allows their members to express themselves freely. See Duhigg (2016).

28. See pages 199–200 of Martin (2002), as well as Sessa (2008) and references therein. See also Grim (2009) and Isaacson (2011). A Macintosh computer is part of the collection of the New York Museum of Modern Art. See the "Apple, Inc." page at https://www.moma.org /collection/works/142218.

29. See Harman et al. (1966).

30. Attributed to Ovid as cited in Sessa (2008), but perhaps apocryphal. A related statement can be traced to Ovid's contemporary Horace (Epistles, Book I, Epistle XIX): "No verses which are written by water-drinkers can please, or be long-lived."

31. See Jarosz et al. (2012). Other studies are discussed on pages 116–117 of Bateson and Martin (2013).

32. See Rees (2010).

33. See Holberton (2005).

34. See Bailey (2010). For other examples of amalgamated art see Kaufmann (2004), Burke (2000), and Bailey (2001).

35. See Scott (2003) for an account of gothic architecture. See Verde (2012) for a historical account of the pointed arch.

36. See pages 160–161 of Csikszentmihalyi (1996).

37. Ibid., 194–295.

38. See page 127 of Simonton (1988).

39. See Hein (1966).

40. Koestler boldly declared that "all decisive advances in the history of scientific thought can be described in terms of mental cross-fertilization between different disciplines." See page 230 of Koestler (1964).

41. See pages 163–165 and 173 of Simonton (1994).

42. See Isaacson (2011) and Appelo (2011).

43. See page 84 of Curtin (1980).

44. See Wilson (1992).

45. See Root-Bernstein et al. (2008).

46. See Simonton (1994) and Csikszentmihalyi (1996).

47. This statement is sometimes also attributed to the novelist Grant Allen.

48. Koestler called this process bisociation. See Koestler (1964).

49. See Chapter 6, page 121 of Koestler (1964).

50. See Chapter 8 of Root-Bernstein and Root-Bernstein (1999), and Schiappa and Van Hee (2012).

51. See page 19 of Arthur (2009).

52. See page 34 of Padel (2008).

53. Several connotations of metaphors used by Aristotle would no longer be associated with metaphors today. See Levin (1982).

54. See pages 145–146 of Root-Bernstein and Root-Bernstein (1999).

55. See page 6 of Pinker (2007).

56. See Tourangeau and Rips (1991).

57. See page 35 of Padel (2008).

58. See page 93 of Csikszentmihalyi (1996).

59. See Simonton (1999).

60. See Guilford (1959), as well as Guilford (1967).

61. Such tests had been used before Guilford's times, but not to assess creativity. See Kent and Rosanoff (1910).

62. These are not the only two dimensions on which responses to a word association test can be scored. Others include flexibility—the ability to create responses that belong in different conceptual categories, such as the use of matches to construct an object or to set fire to it. See page 85 of Simonton (1999) or Kim (2006).

63. See Mednick (1962).

64. Even though the test asks for a single solution to each triplet, it speaks to an important part of creative thinking, the ability to link remote concepts. Its usefulness has been corroborated in validation studies. See page 81 of Simonton (1999) as well as Mednick (1962).

65. See Zeng et al. (2011) or Chapter 6 of Guilford (1967).

66. See Torrance (1966) and Kim (2006). An important class of tests and studies that I do not discuss here revolves around the ability to find problems instead of merely solving problems. See, for example, Csikszentmihalyi and Getzels (1971).

67. See page 4 of Kim (2006).

68. The two limitations of some creativity tests discussed here are not the only ones. Others include that creativity cannot be reduced to a single quantity—it has multiple dimensions—even though some tests aim to compute a single, scalar score. The most general limitation, however, is that the construct creativity itself as a dispositional trait is hard to define. In the words of E. Paul Torrance, the creator of the perhaps most

widely used creativity test: "Creativity defies precise definition. This conclusion does not bother me at all. In fact, I am quite happy with it....However, if we are to study it scientifically, we must have some approximate definition." See Torrance (1988). More generally, test theory knows two fundamental criteria by which to address the question of whether a psychological test "works." The first is to determine reliability, a test's ability to measure a complex quantity such as intelligence or creativity—the psychological term is a *construct*—with similar results across time (test-retest reliability), among different judges (inter-rater reliability), or in other varying contexts. The second criterion is a test's validity, and specifically its *construct validity*, the degree to which a test measures what it aims to measure. Efforts to estimate construct validity often compare the outcome of a test with that of other, independent assessments of creativity, such as creative products. For literature on the reliability and validity of prominent creativity tests see Zeng et al. (2011), Kim (2006), Runco (1992), Torrance (1988), Upmanyu et al. (1996), Mednick (1962), and Gough (1976).

69. See Amabile (1982).

70. The observation that all creativity is ultimately assessed by people is embodied in the widely used Consensual Assessment Technique for creativity, which relies on expert ratings. See Amabile (1982).

71. See Bronson and Merryman (2010).

72. See Torrance (1988) and Plucker (1999).

73. This is why some researchers prefer the term *ideation test* over *creativity test*.

74. The practice of measuring or estimating such distances is a whole lot more sophisticated than I let on, and I do not dwell on this practice, because it is too technical and no universal estimator of distance is agreed upon. Suffice it to say that realistic distance measures themselves are more complex than the simple one I mention in the text, and semantic spaces are not low-dimensional like our three-dimensional continuous space. See Landauer and Dumais (1997). Conventional distance measures in low-dimensional spaces, when applied to concepts, often violate mathematical axioms that distance measures need to fulfill, such that the distance between two objects A and B, $d(A,B)$, is symmetric ($d(A,B)=d(B,A)$), and that it fulfills the so-called triangle inequality $d(A,B) \leq d(A,C) + d(C,B)$. Also, the

relevant spaces need not be continuous but may be discrete, like the spaces of genotypes we encountered earlier. For example, many workers have studied networks of word meanings as graphs—objects that consist of nodes (concepts) that can be connected by edges if their meaning is closely related. Such a graph can be traveled along paths given by its edges. For these reasons, I am tacitly using the notion of a landscape in the most general sense, namely that of a set of objects and a mathematical function from this set onto the real and positive numbers that indicates the appropriateness of an object (such as a combination of concepts) for a particular purpose. As I mentioned in Chapter 7, we still understand little about how our minds represent such collections of objects and how they explore them. See Jones et al. (2011), as well as Griffiths et al. (2007), Landauer and Dumais (1997), Bengio et al. (2003), and Gärdenfors (2000).

75. See Sobel and Rothenberg (1980), as well as Rothenberg (1986).

76. See Rothenberg (1976), Rothenberg (1980), and Rothenberg (2015).

77. See Rothenberg (1995).

78. See Norton (2012) and Chapter 10 of Rothenberg (2015).

79. See Ansburg and Hill (2003).

80. Attributed to Szent-Györgyi on page 14 of IEEE Professional Communication Society (1985). Another pertinent study involved eighty Harvard undergraduate students and showed that the minds of some students ignored previous knowledge to a greater extent. The very same students also scored higher on creativity tests. More than that, they also produced more creative products, such as award-winning works of art. See Carson et al. (2003). The study quantified differences among individuals in latent inhibition, a term from classical conditioning that means that new associations are more difficult to learn for a familiar stimulus than for a new stimulus. Latent inhibition has been detected in many animals, including rats, dogs, and goldfish. See Lubow (1973). Some people show low latent inhibition, and their minds are less capable of ignoring information that the minds of others would filter out because it is likely to be irrelevant for a problem at hand. Latent inhibition is closely related to the phenomenon of negative priming. See Eysenck (1993).

Chapter 9: From Children to Civilizations

1. See "Test-taking" (2013), as well as Lee (2013) and Koo (2014). The exam can be retaken, but those who fail in the first round are still stigmatized because they cannot just use the second, better grade.

2. See Walworth (2015).

3. See Larmer (2014), Zhao (2014), Walworth (2015), as well as Bruni (2015).

4. See the report "PISA 2012 results in focus" available at http://www.oecd.org/pisa/keyfindings/pisa-2012-results.htm.

5. A prominent example is Michael Gove, the British Secretary of State for Education from 2010 to 2014. See Gove (2010).

6. See the IBM survey "Capitalizing on Complexity" available at https://www-01.ibm.com/common/ssi/cgi-bin/ssialias?htmlfid=GBE03297USEN. See also Pappano (2014).

7. Cited on page 305 of Runco (2014).

8. See Table 5 of Bassok and Rorem (2014).

9. See Chapter 2 of Zhao (2014).

10. See pages 40–41 of Zhao (2014).

11. See page 139 of Zhao (2014), Koo (2014), as well as Zhao and Gearin (2016).

12. See Kim (2011), as well as Bronson and Merryman (2010).

13. See Niu and Sternberg (2001, 2003).

14. See Marcon (2002).

15. See Kohn (2015), as well as Rich (2015) and Marcon (2002).

16. See Ruef (2005) and the website *The Private Eye* at http://www.the-private-eye.com/.

17. See Arieff (2015) and her website *Project H* at http://www.projecthdesign.org/.

18. See Garaigordobil (2006). The children in this study were ten and eleven years old, but the benefits of play are evident even earlier. Elementary school boys, for example, who regularly engage in rough-housing in the schoolyard are better social problem solvers, and toddlers given the opportunity to play with toy blocks (instead of watching TV) develop better language skills. See Pellegrini (1988) and Christakis et al. (2007).

19. See Scott et al. (2004), as well as Runco (2001) and Niu and Sternberg (2003).

20. See, for example, Kamenetz (2015) and Grant (2014). And just as there are alternatives to assessing students, there are alternatives to assessing teachers. See Nocera (2015).

21. See Hiss and Franks (2014), and also National Association for College Admission Counseling (2008).

22. See Gaugler et al. (1987), as well as Grant (2014).

23. I focus here on college admission testing, but just as insidious is the problem of high-stakes testing throughout the school year, for example to determine teacher effectiveness, because frequent tests continually distract from broader teaching goals.

24. It is no coincidence that the Finnish education system, one of the world's best-ranked school systems that does not run on high-stakes standardized tests, cultivates the autonomy of teachers and schools. Other contributors to its success include a high social standing of teachers and high-quality teacher training. See Sahlberg (2015).

25. See Amabile (1985). See also Hennessey and Amabile (1998).

26. The opposite is also true: simply thinking about intrinsic reasons for pursuing an activity may be sufficient to enhance creativity in that activity. More generally, only some kinds of extrinsic motivation are detrimental for creativity, especially those that make a person feel controlled or as though they have lost autonomy in performing a task. See Collins and Amabile (1999).

27. See page 328 of Csikszentmihalyi (1996).

28. See page 335 of Csikszentmihalyi (1996).

29. Cited on page 158 of Simonton (1994).

30. See Kim (2006), Westby and Dawson (1995), Torrance (1972), and page 173 of Runco (2014).

Some older studies suggest that even their own parents view creative children unfavorably. See Raina (1975).

31. See Pomerantz et al. (2014).

32. Name changed.

33. I also remember being shocked at first by the limited biological knowledge of the American biology students I helped teach, as compared to European students. This knowledge gap reflects a difference between Central European and American high schools that is often

mentioned as a symptom of America's looming decline. Remarkably, it was already noted by a 1916 French visitor to the United States, long before the United States would become a global leader in science and technology. See Rosenberg and Nelson (1994). Maximizing the amount of information crammed into a young person's head is clearly not the most important thing an education achieves.

34. See pages 170–171 of Ramon y Cajal (1951).

35. See Wuchty et al. (2007). It is important to not equate influence with quality, because some of the best research remains obscure for a long time. The best-known example is the work of Gregor Mendel in the nineteenth century, which remained without influence for half a century, but eventually helped trigger the genetics revolution of the twentieth century. Additionally, not all citations of a scientist's work reflect intellectual debt. Some controversial works, for example, attract negative citations that disparage the work. This is why it is not advisable to evaluate the work of individual scientists based on citations alone, even though citation patterns can be helpful to identify broad historical trends. See also Adler et al. (2009).

36. Examples of two such highly influential publications include Newman et al. (2001), as well as West et al. (1997).

37. See page 238 in Bush (1945), which is a reprint of the original document by the US government printing office.

38. See the National Science Foundation website, the "The Nobel Prizes" page, at https://www.nsf.gov/news/special_reports/nobelprizes/.

39. See Figures 1, 5, and 13 of National Institutes of Health (2012), as well as Alberts et al. (2014).

40. The funding rates of full proposals lie in the neighborhood of 20 percent, but submission of a full proposal is preceded by a mandatory pre-proposal in several programs of the NSF Directorate for Biological Sciences, with acceptance rates that are similarly low, resulting in very low overall funding rates. See National Science Foundation (2014) and Appendix 2 therein.

41. See Adler et al. (2009).

42. See Lee et al. (2013).

43. See Alberts et al. (2014).

44. These would be the select few researchers who already have made the cut imposed by academic faculty search committees, which

examine, compare, and discuss the records of many applicants in detail, an arduous process for which there is no shortcut. It is also worth pointing out that young US researchers do usually receive some noncompetitive "start-up" funding from their universities, but that funding is intended to equip and start a research laboratory. Because it runs out after a few years, it is no long-term solution to avoiding hypercompetition.

45. To be sure, some US institutions like the Howard Hughes Medical Institute (HHMI) effectively use a similar strategy by funding individuals rather than projects. Consistent with the Darwinian perspective on creativity, the strategy leads to more flops, but also to bigger breakthroughs, as a comparison between HHMI- and NIH-funded investigators shows. See Azoulay et al. (2011). Being available only to a small and well-established elite in limited fields like biomedicine, such funding is a drop in the proverbial bucket.

46. For comparative research statistics, see State Secretariat for Education and Research (2011). A caveat is that impact statistics fluctuate over the years, but Swiss science remains strong even when such fluctuations are taken into account. Other reasons include good public schools and high investment in R&D. (Switzerland invests 3.4 percent of its gross domestic product into R&D, more than the United States at 2.7 percent, according to 2015 OECD statistics available at https://data.oecd.org/rd/gross-domestic-spending-on-r-d.htm.) Also important are low corruption and little nepotism in academic hiring, which plagues academia in some developed countries.

47. See Zappe (2013), as well as Rosenberg and Nelson (1994) and Porter (2015).

48. Additional benefits provided by universities include not just workforce training, but also a knowledge base that is necessary to take advantage of the latest science. See Pavitt (2001), as well as Callon (1994), Salter and Martin (2001), and Rosenberg and Nelson (1994).

49. For the long breath needed to commercialize fundamental discoveries, see Rosenberg and Nelson (1994), as well as Pavitt (2001) and Zappe (2013).

50. See Gertner (2012a) and Gertner (2012b).

51. Albeit very visible, they are perhaps exceptions to a rule of declining research by large corporations. See Arora et al. (2015).

52. See Amabile et al. (2002).

53. And when creative people get busy, they usually work on multiple projects, another way of linking different domains of knowledge. See Schwartz (2013) and page 113 of Sawyer (2013).

54. See Schwartz (2013).

55. See Amabile (1998).

56. See Kelley (2001) on IDEO. Psychological research also holds some surprising lessons about how diverse teams operate best. One of them is about the time-honored device of brainstorming, which is not necessarily the best way to collect diverse ideas. The simple reason is that it is very difficult to completely turn off the evaluation of ideas in a group setting, perhaps because such evaluation can take exceedingly subtle forms. It may sometimes be better if individual group members think about a problem at hand and then compare notes and discuss their candidate solutions. See pages 158–159 and 188–189 of Runco (2014).

57. See Amabile (1998).

58. See Amabile et al. (2002).

59. See Frese and Keith (2015).

60. See Slack (2002), as well as Osepchuk (1984).

61. The reasons why this class of innovators could emerge are complex, but they include a participatory form of government and strong property rights. See Acemoglu and Robinson (2012), as well as Rosen (2010).

62. See page 161 of Zhao (2014).

63. See Normile (2015).

64. Other pertinent statistics are summarized in Bruni (2017). On the relationship between mobility and multicultural experiences on creativity, see Maddux et al. (2010), as well as Maddux and Galinsky (2009) and Leung et al. (2008). An analogous link between racial diversity and creativity is pointed out by Velasquez-Manoff (2017). For the influence of immigrants, see pages 122–125 of Simonton (1999).

65. See Alberts et al. (2014).

66. See Gustin (2013). Such visa policies are unfortunately also subject to abuse that needs to be prevented. A case in point is an incident at Walt Disney World, where skilled IT workers were fired only to be replaced with cheaper foreign immigrants of comparable skill. See Preston (2015a), as well as Preston (2015b).

67. See Godart et al. (2015).

68. The pattern holds not just in the West, but also in Islamic and Indian civilizations. See Simonton (1975, 1990).

69. See Simonton (1975, 1990). Simonton cites China as a potential exception because of its greater cultural homogeneity. It is intriguing that minority views can enhance divergent thinking also in psychological experiments. See Nemeth and Kwan (1987).

70. See Simonton (1997).

71. See Wagner and Jonkers (2017), as well as Sugimoto et al. (2017).

72. See page 48 of McArdle (2014).

73. Ibid., 251.

74. Within six years of being founded in 2009, FailCon has been exported from its birthplace in San Francisco to half a dozen countries. Clearly, entrepreneurs everywhere are hungry to keep going. See Martin (2014), as well as Stewart (2015), and pages 48–51 of McArdle (2014). See Birrane (2017) for Fuck Up Nights.

75. Specifically, I am referring to Chapter 7 of Title 11 of the US Code (U.S.C.). Other forms of bankruptcy, such as Chapter 13, require debtors to pay back more of their debts. Historical accident rather than political farsightedness may deserve the credit for this mechanism of debt relief. That accident came in the form of indebted farmers who lobbied the Senate for debt relief more than a century ago. See page 249 of McArdle (2014).

76. See "Morally bankrupt" (2005).

77. See Piketty (2014).

78. The interdependent self is not strong just in Asian societies, but also in Africa and Latin America. See page 228 of Markus and Kitayama (1991). Among all Western cultures, the independent self appears to be strongest in US Americans. See pages 74–75 of Henrich et al. (2010). The Japanese proverb is taken from Markus and Kitayama (1991).

79. Other deep roots may exist as well, such as a postulated if controversial connection to wheat and rice agriculture. See Talhelm et al. (2014).

80. See pages 15–16 of Cheng (1998).

81. For the limited impact of Chinese inventions on Chinese culture see pages 77–80 of Zhao (2014), as well as pages 252–253 of Runco (2014). When threatened by British gunboats in the nineteenth century, China aimed to buy Western weapons technology without "contaminating" its culture with Western values. Unfortunately "gunboats and steel mills bring their own philosophy with them," as historians John King Fairbank and Merle Goldman observe (cited on page 80 of Zhao [2014]).

82. The Yale–Beijing study's observations are echoed by another study where Chinese graduate students scored lower on a creativity test. See Zha et al. (2006), as well as Niu and Sternberg (2001, 2003).

83. See Aviram and Milgram (1977).

84. See Amabile (1998).

85. See page 16 of Cheng (1998).

86. See Hennessey and Amabile (2010).

Epilogue: More than Metaphors

1. See de Visser and Krug (2014).

Bibliography

Acemoglu, D., and Robinson, J.A. 2012. *Why Nations Fail*. Crown Publishers, New York.

Adams, T. 2010. "David Cope: 'You pushed the button and out came hundreds and thousands of sonatas.'" *Guardian*. July 7.

Adler, R., Ewing, J., Taylor, P., and Hall, P.G. 2009. "A report from the International Mathematical Union (IMU) in cooperation with the International Council of Industrial and Applied Mathematics (ICIAM) and the Institute of Mathematical Statistics (IMS)." *Statistical Science* **24**, 1.

Aguilar-Rodriguez, J., Payne, J.L., and Wagner, A. 2017. "1000 empirical adaptive landscapes and their navigability." *Nature Ecology and Evolution* **1**, 45.

Alberts, B., Kirschner, M.W., Tilghman, S., and Varmus, H. 2014. "Rescuing US biomedical research from its systemic flaws." *Proceedings of the National Academy of Sciences of the United States of America* **111**, 5773.

Alvarez, G., Ceballos, F.C., and Quinteiro, C. 2009. "The role of inbreeding in the extinction of a European royal dynasty." *PLoS ONE* **4**.

Amabile, T.M. 1982. "Social psychology of creativity—a consensual assessment technique." *Journal of Personality and Social Psychology* **43**, 997.

———. 1985. "Motivation and creativity—effects of motivational orientation on creative writers." *Journal of Personality and Social Psychology* **48**, 393.

———. 1998. "How to kill creativity." *Harvard Business Review* **76**, 76.

Amabile, T.M., Hadley, C.N., and Kramer, S.J. 2002. "Creativity under the gun." *Harvard Business Review* **80**, 52.

Anderson, T.M., vonHoldt, B.M., Candille, S.I., Musiani, M., Greco, C., Stahler, D.R., Smith, D.W., Padhukasahasram, B., Randi, E., Leonard, J.A., Bustamante, C.D., Ostrander, E.A., Tang, H., Wayne, R.K., and Barsh, G.S. 2009. "Molecular and evolutionary history of melanism in North American gray wolves." *Science* **323**, 1339.

Ansburg, P.I., and Hill, K. 2003. "Creative and analytic thinkers differ in their use of attentional resources." *Personality and Individual Differences* **34**, 1141.

Appelo, T. 2011. "How a calligraphy pen rewrote Steve Jobs' life." *Hollywood Reporter*. www.hollywoodreporter.com (Retrieved on August 20, 2014).

Arieff, A. 2015. "Learning through tinkering." *New York Times*. April 3.

Arnold, M.L., Bulger, M.R., Burke, J.M., Hempel, A.L., and Williams, J.H. 1999. "Natural hybridization: How low can you go and still be important?" *Ecology* **80**, 371.

Arnold, M.L., and Hodges, S.A. 1995. "Are natural hybrids fit or unfit relative to their parents?" *Trends in Ecology & Evolution* **10**, 67.

Arnold, M.L., and Kunte, K. 2017. "Adaptive genetic exchange: A tangled history of admixture and evolutionary innovation." *Trends in Ecology & Evolution* **32**, 601.

Aronson, H., Royer, W., and Hendrickson, W. 1994. "Quantification of tertiary structural conservation despite primary sequence drift in the globin fold." *Protein Science* **3**, 1706.

Arora, A., Belenzon, S., and Patacconi, A. 2015. "Killing the golden goose? The decline of science in corporate R&D." (NBER working paper no. 20902.) National Bureau of Economic Research, Cambridge, MA.

Arthur, W.B. 2009. *The Nature of Technology: What It Is and How It Evolves*. Free Press, New York.

Aviram, A., and Milgram, R.M. 1977. "Dogmatism, locus of control, and creativity in children educated in the Soviet Union, the United States, and Israel." *Psychological Reports* **40**, 27.

Azoulay, P., Graff Zivin, J.S., and Manso, G. 2011. "Incentives and creativity: Evidence from the academic life sciences." *The RAND Journal of Economics* **42**, 527.

Badis, G., Berger, M.F., Philippakis, A.A., Talukder, S., Gehrke, A.R., Jaeger, S.A., Chan, E.T., Metzler, G., Vedenko, A., Chen, X., Kuznetsov, H., Wang, C.-F., Coburn, D., Newburger, D.E., Morris, Q., Hughes, T.R., and Bulyk, M.L. 2009. "Diversity and complexity in DNA recognition by transcription factors." *Science* **324**, 1720.

Bailey, G.A. 2001. *Art on the Jesuit Missions in Asia and Latin America, 1542–1773.* University of Toronto Press, Toronto.

———. 2010. *The Andean Hybrid Baroque.* University of Notre Dame Press, Notre Dame, IN.

Baird, B., Smallwood, J., Mrazek, M.D., Kam, J.W.Y., Franklin, M.S., and Schooler, J.W. 2012. "Inspired by distraction: Mind wandering facilitates creative incubation." *Psychological Science* **23**, 1117.

Baker, B.M., and Ayechew, M.A. 2003. "A genetic algorithm for the vehicle routing problem." *Computers & Operations Research* **30**, 787.

Ball, P. 2012. "Iamus, classical music's computer composer, live from Malaga." *Guardian.* July 1.

Banzhaf, W., and Leier, A. 2006. "Evolution on neutral networks in genetic programming." In *Genetic Programming Theory, and Practice III, Genetic Programming Vol. 9,* eds. T. Yu, R. Riolo, and B. Worzel, p. 207. Springer, Boston, MA.

Baror, S., and Bar, M. 2016. "Associative activation and its relation to exploration and exploitation in the brain." *Psychological Science* **27**, 776.

Bassok, D., and Rorem, A. 2014. "Is kindergarten the new first grade? The changing nature of kindergarten in the age of accountability." *EdPolicyWorks Working Paper Series, No. 20.* http://curry.virginia .edu/uploads/resourceLibrary/20_Bassok_Is_Kindergarten_The _New_First_Grade.pdf.

Bateson, P., and Martin, P. 2013. *Play, Playfulness, Creativity and Innovation.* Cambridge University Press, Cambridge, UK.

Beech, A., and Claridge, G. 1987. "Individual differences in negative priming—relations with schizotypal personality traits." *British Journal of Psychology* **78**, 349.

Bell, M.A. 2012. "Adaptive landscapes, evolution, and the fossil record." In *The Adaptive Landscape in Evolutionary Biology*, eds. E.I. Svensson and R. Calsbeek, p. 243. Oxford University Press, Oxford, UK.

Bengio, Y., Ducharme, R., Vincent, P., and Jauvin, C. 2003. "A neural probabilistic language model." *Journal of Machine Learning Research* **3**, 1137.

Benson, W.W. 1972. "Natural selection for Mullerian mimicry in Heliconius erato in Costa Rica." *Science* **176**, 936.

Berne, O., and Tielens, A. 2012. "Formation of buckminsterfullerene (C-60) in interstellar space." *Proceedings of the National Academy of Sciences of the United States of America* **109**, 401.

Berry, R.S. 1993. "Potential surfaces and dynamics—what clusters tell us." *Chemical Reviews* **93**, 2379.

Bershtein, S., Goldin, K., and Tawfik, D.S. 2008. "Intense neutral drifts yield robust and evolvable consensus proteins." *Journal of Molecular Biology* **379**, 1029.

Biery, M.E. 2014. "U.S. trucking companies deliver sales, profit gains." *Forbes*. February 20. http://www.forbes.com/sites/sageworks/2014/02/20/sales-profit-trends-trucking-companies/.

Birrane, A. 2017. "Yes, you should tell everyone about your failures." *BBC Capital*. March 13. http://www.bbc.com/capital/story/20170312-yes-you-should-tell-everyone-about-your-failures.

Bronson, P., and Merryman, A. 2010. "The creativity crisis." *Newsweek*. July 10. https://www.newsweek.com/creativity-crisis-74665.

Brower, A.V.Z. 1994. "Rapid morphological radiation and convergence among races of the butterfly *Heliconius erato* inferred from patterns of mitochondrial DNA evolution." *Proceedings of the National Academy of Sciences of the United States of America* **91**, 6491.

———. 2013. "Introgression of wing pattern alleles and speciation via homoploid hybridization in Heliconius butterflies: A review of evidence from the genome." *Proceedings of the Royal Society B-Biological Sciences* **280**.

Brown, K.S.J. 1981. "The biology of Heliconius and related genera." *Annual Review of Entomology* **26**, 427.

Bruni, F. 2015. "Best, brightest—and saddest?" *New York Times*. April 11.

———. 2017. "Want geniuses? Welcome immigrants." *New York Times*. September 23.

Burke, P. 2000. *Kultureller Austausch*. Suhrkamp, Frankfurt am Main.

Bush, V. 1945. "Science: The endless frontier." *Transactions of the Kansas Academy of Science* **48**, 231.

Bushman, F. 2002. *Lateral DNA Transfer: Mechanisms and Consequences*. Cold Spring Harbor University Press, Cold Spring Harbor, NY.

Callon, M. 1994. "Is science a public good—5th Mullin lecture, Virginia Polytechnic Institute, 23 March 1993." *Science Technology & Human Values* **19**, 395.

Cameron, E.Z., Linklater, W.L., Stafford, K.J., and Minot, E.O. 2008. "Maternal investment results in better foal condition through increased play behaviour in horses." *Animal Behaviour* **76**, 1511.

Cami, J., Bernard-Salas, J., Peeters, E., and Malek, S.E. 2010. "Detection of C-60 and C-70 in a young planetary nebula." *Science* **329**, 1180.

Campbell, C.D., and Eichler, E.E. 2013. "Properties and rates of germline mutations in humans." *Trends in Genetics* **29**, 575.

Campbell, D.T. 1960. "Blind variation and selective retention in creative thought as in other knowledge processes." *Psychological Review* **67**, 380.

Campbell, E., Holz, M., Gerlich, D., and Maier, J. 2015. "Laboratory confirmation of C60+ as the carrier of two diffuse interstellar bands." *Nature* **523**, 322.

Carbone, C., and Gittleman, J.L. 2002. "A common rule for the scaling of carnivore density." *Science* **295**, 2273.

Caro, T.M. 1995. "Short-term costs and correlates of play in Cheetahs." *Animal Behaviour* **49**, 333.

Carson, S.H., Peterson, J.B., and Higgins, D.M. 2003. "Decreased latent inhibition is associated with increased creative achievement in high-functioning individuals." *Journal of Personality and Social Psychology* **85**, 499.

Cartwright, J. 2012. "Pico-gold clusters break catalysis record." *Chemistry World*. December 14. http://www.rsc.org/chemistryworld/2012/12/nano-gold-catalyst-record-breaking.

Chamberlain, J.A. 1976. "Flow patterns and drag coefficents of cephalopod shells." *Palaeontology (Oxford)* **19**, 539.

———. 1981. "Hydromechanical design of fossil cephalopods." In *The Ammonoidea: The evolution, classification, mode of life and geological usefulness of a major fossil group*, eds. M.R. House and J.R. Senior, p. 289. Academic Press, London.

Charlesworth, D., and Willis, J.H. 2009. "The genetics of inbreeding depression." *Nature Reviews Genetics* **10**, 783.

Cheng, K.-M. 1998. "Can education values be borrowed? Looking into cultural differences." *Peabody Journal of Education* **73**, 11.

Chimpanzee Sequencing and Analysis Consortium. 2005. "Initial sequence of the chimpanzee genome and comparison with the human genome." *Nature* **437**, 69.

Chipp, H.B. 1988. *Picasso's Guernica*. University of California Press, Berkeley, CA.

Christakis, D.A., Zimmerman, F.J., and Garrison, M.M. 2007. "Effect of block play on language acquisition and attention in toddlers— A pilot randomized controlled trial." *Archives of Pediatrics & Adolescent Medicine* **161**, 967.

Christoff, K. 2012. "Undirected thought: Neural determinants and correlates." *Brain Research* **1428**, 51.

Clark, R. 2013. *J.B.S. The Life and Work of J.B.S. Haldane*. Bloomsbury Reader, London, UK.

Clerwall, C. 2014. "Enter the robot journalist. Users' perceptions of automated content." *Journalism Practice* **8**, 519.

Collins, M.A., and Amabile, T.M. 1999. "Motivation and creativity." In *Handbook of creativity*, ed. R.J. Sternberg, p. 297. Cambridge University Press, Cambridge, UK.

Coltman, D.W., Pilkington, J.G., Smith, J.A., and Pemberton, J.M. 1999. "Parasite-mediated selection against inbred Soay sheep in a free-living, island population." *Evolution* **53**, 1259.

Constine, J. 2015. "Need music for a video? Jukedeck's AI composer makes cheap, custom soundtracks." *TechCrunch*. December 7. https://techcrunch.com/2015/12/07/jukedeck/.

Cook, W.J. 2012. *In Pursuit of the Traveling Salesman*. Princeton University Press, Princeton, NJ.

Cope, D. 1991. "Recombinant music—using the computer to explore musical style." *Computer* **24**, 22.

Copley, S.D., Rokicki, J., Turner, P., Daligault, H., Nolan, M., and Land, M. 2012. "The whole genome sequence of *Sphingobium chlorophenolicum L-1*: Insights into the evolution of the pentachlorophenol degradation pathway." *Genome Biology and Evolution* **4**, 184.

Corma, A., Concepcion, P., Boronat, M., Sabater, M.J., Navas, J., Yacaman, M.J., Larios, E., Posadas, A., Lopez-Quintela, M.A., Buceta, D., Mendoza, E., Guilera, G., and Mayoral, A. 2013. "Exceptional oxidation activity with size-controlled supported gold clusters of low atomicity." *Nature Chemistry* **5**, 775.

Coyne, J. 2005. "The faith that dare not speak its name: The case against intelligent design." *New Republic*. August 22–29.

Crameri, A., Dawes, G., Rodriguez, E., Silver, S., and Stemmer, W. 1997. "Molecular evolution of an arsenate detoxification pathway DNA shuffling." *Nature Biotechnology* **15**, 436.

Crameri, A., Raillard, S., Bermudez, E., and Stemmer, W. 1998. "DNA shuffling of a family of genes from diverse species accelerates directed evolution." *Nature* **391**, 288.

Csikszentmihalyi, M. 1996. *Creativity: The Psychology of Discovery and Invention*. Harper Collins, New York.

Csikszentmihalyi, M., and Getzels, J.W. 1971. "Discovery-oriented behavior and the originality of creative products: A study with artists." *Journal of Personality and Social Psychology* **19**, 47.

Curtin, D.W. 1980. *The Aesthetic Dimension of Science*. Philosophical Library, New York.

Dantzig, G.B. 1963. *Linear Programming and Extensions*. Princeton University Press, Princeton, NJ.

Darwin, C. 1859. *On the Origin of Species by Means of Natural Selection, or the Preservation of Favored Races in the Struggle for Life (1st ed.)*. John Murray, London.

———. 1868. *Animals and Plants Under Domestication*. John Murray, London.

Dasgupta, S. 2004. "Is creativity a Darwinian process?" *Creativity Research Journal* **16**, 403.

Dawkins, R. 1976. *The Selfish Gene*. Oxford University Press, New York.

de Visser, J.A.G.M., and Krug, J. 2014. "Empirical fitness landscapes and the predictability of evolution." *Nature Reviews Genetics* **15**, 480.

Dehaene, S. 2014. *Consciousness and the Brain*. Penguin, New York.

Dietrich, M.R., and Skipper Jr., R.A. 2012. "A shifting terrain: A brief history of the adaptive landscape." In *The Adaptive Landscape in Evolutionary Biology*, eds. E.I. Svensson and R. Calsbeek, p. 3. Oxford University Press, Oxford, UK.

Drummond, D.A., Silberg, J.J., Meyer, M.M., Wilke, C.O., and Arnold, F.H. 2005. "On the conservative nature of intragenic recombination." *Proceedings of the National Academy of Sciences of the United States of America* **102**, 5380.

Duhigg, C. 2016. "What Google learned from its quest to build the perfect team." *New York Times*. February 28.

Eiberg, H., Troelsen, J., Nielsen, M., Mikkelsen, A., Mengel-From, J., Kjaer, K.W., and Hansen, L. 2008. "Blue eye color in humans may be caused by a perfectly associated founder mutation in a regulatory element located within the HERC2 gene inhibiting OCA2 expression." *Human Genetics* **123**, 177.

Eyre-Walker, A., and Keightley, P.D. 2007. The distribution of fitness effects of new mutations. *Nature Reviews Genetics* **8**, 610.

Eysenck, H.J. 1993. "Creativity and personality: Suggestions for a theory." *Psychological Inquiry* **4**, 147.

Fagen, R., and Fagen, J. 2009. "Play behaviour and multi-year juvenile survival in free-ranging brown bears, Ursus arctos." *Evolutionary Ecology Research* **11**, 1.

Fernandez, J.D., and Vico, F. 2013. "AI methods in algorithmic composition: A comprehensive survey." *Journal of Artificial Intelligence Research* **48**, 513.

Flot, J.F., Hespeels, B., Li, X., Noel, B., Arkhipova, I., Danchin, E.G.J., Hejnol, A., Henrissat, B., Koszul, R., Aury, J.M., Barbe, V., Barthelemy, R.M., Bast, J., Bazykin, G.A., Chabrol, O., Couloux, A., Da Rocha, M., Da Silva, C., Gladyshev, E., Gouret, P., Hallatschek, O., Hecox-Lea, B., Labadie, K., Lejeune, B., Piskurek, O., Poulain, J., Rodriguez, F., Ryan, J.F., Vakhrusheva, O.A., Wajnberg, E., Wirth, B., Yushenova, I., Kellis, M., Kondrashov, A.S., Welch,

D.B.M., Pontarotti, P., Weissenbach, J., Wincker, P., Jaillon, O., and Van Doninck, K. 2013. "Genomic evidence for ameiotic evolution in the bdelloid rotifer Adineta vaga." *Nature* **500**, 453.

Fraser, C., Hanage, W.P., and Spratt, B.G. 2007. "Recombination and the nature of bacterial speciation." *Science* **315**, 476.

Freeman, S., and Herron, J.C. 2007. *Evolution (4th ed)*. Pearson, San Francisco.

Frese, M., and Keith, N. 2015. "Action errors, error management, and learning in organizations." *Annual Review of Psychology* **66**, 661.

Futuyma, D.J. 2009. *Evolution*. Sinauer, Sunderland, MA.

Garaigordobil, M. 2006. "Intervention in creativity with children aged 10 and 11 years: Impact of a play program on verbal and graphic-figural creativity." *Creativity Research Journal* **18**, 329.

Garcia-Hernandez, D.A., Manchado, A., Garcia-Lario, P., Stanghellini, L., Villaver, E., Shaw, R.A., Szczerba, R., and Perea-Calderon, J.V. 2010. "Formation of fullerenes in H-containing planetary nebulae." *Astrophysical Journal Letters* **724**, L39.

Gärdenfors, P. 2000. *Conceptual Spaces: The Geometry of Thought*. MIT Press, Cambridge, MA.

Gaugler, B.B., Rosenthal, D.B., Thornton III, G.C., and Bentson, C. 1987. "Meta-analysis of assessment center validity." *Journal of Applied Psychology* **72**, 493.

Gavrilets, S. 1997. "Evolution and speciation on holey adaptive landscapes." *Trends in Ecology & Evolution* **12**, 307.

Geist, D.J., Snell, H., Snell, H., Goddard, C., and Kurz, M.D. 2014. "A paleogeographic model of the Galápagos Islands and biogeographical and evolutionary implications." In *The Galápagos: A Natural Laboratory for the Earth Sciences*, eds. K.S. Harpp, E. Mittelstaedt, N. d'Ozouville, and D.W. Graham, p. 145. American Geophysical Union, Washington, DC.

Gelvin, S.B. 2003. "Agobacterium-mediated plant transformation: The biology behind the 'gene-jockeying' tool." *Microbiology and Molecular Biology Reviews* **67**, 16.

Gerst, C. 2013. *Buckminster Fuller: Poet of Geometry*. Overcup Press, Portland, OR.

Gertner, J. 2012a. *The Idea Factory: Bell Labs and the Great Age of American Innovation*. Penguin, New York.

———. 2012b. "True innovation." *New York Times*. February 25.

Gilbert, W. 1978. "Why genes in pieces?" *Nature* **271**, 501.

Glover, F., and Kochenberger, G.A. 2003. *Handbook of Metaheuristics*. Kluwer Academic Publishers, New York.

Godart, F.C., Maddux, W.W., Shipilov, A.V., and Galinsky, A.D. 2015. "Fashion with a foreign flair: Professional experiences abroad facilitate the creative innovations of organizations." *Academy of Management Journal* **58**, 195.

Gough, H.G. 1976. "Studying creativity by means of word-association tests." *Journal of Applied Psychology* **61**, 348.

Gove, M. 2010. "Michael Gove: My revolution for culture in the classroom." *Telegraph*. December 28.

Grant, A. 2014. "Throw out the college application system." *New York Times*. October 4.

Grant, P.R. 1998. "Patterns on islands and microevolution." In *Evolution on Islands*, ed. P.R. Grant, p. 1. Oxford University Press, Oxford, UK.

Grant, P.R., and Grant, B.R. 2009. "The secondary contact phase of allopatric speciation in Darwin's finches." *Proceedings of the National Academy of Sciences of the United States of America* **106**, 20141.

Graveley, B.R. 2001. "Alternative splicing: Increasing diversity in the proteomic world." *Trends in Genetics* **17**, 100.

Griffiths, A., Wessler, S., Lewontin, R., Gelbart, W., Suzuki, D., and Miller, J. 2004. *An Introduction to Genetic Analysis*. Freeman, New York.

Griffiths, T.L., Steyvers, M., and Tenenbaum, J.B. 2007. "Topics in semantic representation." *Psychological Review* **114**, 211.

Grim, R. 2009. "Read the never-before-published letter from LSD-inventor Albert Hofmann to Apple CEO Steve Jobs." *Huffington Post*. August 8.

Guilford, J.P. 1959. "Three faces of intellect." *American Psychologist* **14**, 469.

———. 1967. *The Nature of Human Intelligence*. McGraw-Hill, New York.

Gustin, S. 2013. "Why Mark Zuckerberg is pushing for immigration reform." *Time*. April 13.

Hadamard, J. 1945. *The Psychology of Invention in the Mathematical Field*. Dover, New York.

Haffer, J. 1969. "Speciation in Amazonian forest birds." *Science* **165**, 131.

Haldane, J.B.S. 1924. "A mathematical theory of natural and artificial selection. Part I." *Transactions of the Cambridge Philosophical Society* **23**, 19.

Harcourt, R. 1991. "Survivorship costs of play in the South-American fur seal." *Animal Behaviour* **42**, 509.

Hardison, R. 1999. "The evolution of hemoglobin." *American Scientist* **87**, 126.

Harman, W.W., McKim, R.H., Mogar, R.E., Fadiman, J., and Stolaroff, M.J. 1966. "Psychedelic agents in creative problem solving—a pilot study." *Psychological Reports* **19**, 211.

Hartl, D.L., and Clark, A.G. 2007. *Principles of Population Genetics*. Sinauer Associates, Sunderland, MA.

Hawass, Z., Gad, Y.Z., Ismail, S., Khairat, R., Fathalla, D., Hasan, N., Ahmed, A., Elleithy, H., Ball, M., Gaballah, F., Wasef, S., Fateen, M., Amer, H., Gostner, P., Selim, A., Zink, A., and Pusch, C.M. 2010. "Ancestry and pathology in King Tutankhamun's family." *Journal of the American Medical Association* **303**, 638.

Hay-Roe, M.M., and Nation, J. 2007. "Spectrum of cyanide toxicity and allocation in Heliconius erato and Passiflora host plants." *Journal of Chemical Ecology* **33**, 319.

Hayden, E., Ferrada, E., and Wagner, A. 2011. "Cryptic genetic variation promotes rapid evolutionary adaptation in an RNA enzyme." *Nature* **474**, 92.

Hayden, E.J., and Wagner, A. 2012. "Environmental change exposes beneficial epistatic interactions in a catalytic RNA." *Proceedings of the Royal Society B-Biological Sciences* **279**, 3418.

Hein, G.E. 1966. "Kekule and the architecture of molecules." *Advances in Chemistry Series* 1.

Henig, R.M. 2008. "Taking play seriously." *New York Times*. February 17.

Hennessey, B.A., and Amabile, T.M. 1998. "Reward, intrinsic motivation, and creativity." *American Psychologist* **53**, 674.

————. 2010. "Creativity." *Annual Review of Psychology* **61**, 569.

Henrich, J., Heine, S.J., and Norenzayan, A. 2010. "The weirdest people in the world?" *Behavioral and Brain Sciences* **33**, 61.

Hernando, L., Mendiburu, A., and Lozano, J.A. 2013. "An evaluation of methods for estimating the number of local optima in combinatorial optimization problems." *Evolutionary Computation* **21**, 625.

Hiraishi, A. 2008. "Biodiversity of dehalorespiring bacteria with special emphasis on polychlorinated biphenyl/dioxin dechlorinators." *Microbes and Environments* **23**, 1.

Hiss, W.C., and Franks, V.W. 2014. "Defining promise: Optional standardized testing policies in American college and university admissions." *Report of the National Association for College Admission Counseling (NACAC)*. http://www.nacacnet.org/research/research-data/nacac-research/Documents/Defining Promise.pdf.

Holberton, P. 2005. "Bellini and the East." National Gallery Company Limited, London.

Holland, J.H. 1975. *Adaptation in Natural and Artificial Systems*. University of Michigan Press, Ann Arbor.

Holland, O.A. 1987. "Schnittebenenverfahren für travelling-salesman und verwandte Probleme." Universität Bonn, Bonn, Germany.

Hornby, G.S., Lohn, J.D., and Linden, D.S. 2011. "Computer-automated evolution of an X-band antenna for NASA's space technology 5 mission." *Evolutionary Computation* **19**, 1.

Hosseini, S.-R., Martin, O.C., and Wagner, A. 2016. "Phenotypic innovation through recombination in genome-scale metabolic networks." *Proceedings of the Royal Society B-Biological Sciences* **283**, 2016.1536.

IEEE Professional Communication Society. 1985. "Bridging the present and the future: IEEE Professional Communication Society conference record, Williamsburg, Virginia, October 16–18, 1985." Institute of Electrical and Electronics Engineers, New York.

Isaacson, W. 2011. *Steve Jobs*. Simon and Schuster, New York.

Jackson, J.D., and Balota, D.A. 2012. "Mind-wandering in younger and older adults: Converging evidence from the Sustained Attention to Response Task and reading for comprehension." *Psychology and Aging* **27**, 106.

James, W. 1880. "Great men, great thoughts, and the environment." *Atlantic Monthly* **46**, 441.

Jarosz, A.F., Colflesh, G.J.H., and Wiley, J. 2012. "Uncorking the muse: Alcohol intoxication facilitates creative problem solving." *Consciousness and Cognition* **21**, 487.

Jimenez, J.I., Xulvi-Brunet, R., Campbell, G.W., Turk-MacLeod, R., and Chen, I.A. 2013. "Comprehensive experimental fitness landscape and evolutionary network for small RNA." *Proceedings of the National Academy of Sciences of the United States of America* **110**, 14984.

John-Steiner, V. 1997. *Notebooks of the Mind: Explorations of Thinking.* Oxford University Press, Oxford, UK.

Johnson, G. 1997. "Undiscovered Bach? No, a computer wrote it." *New York Times.* November 11.

Jones, M.N., Gruenenfelder, T.M., and Recchia, G. 2011. "In defense of spatial models of lexical semantics." In *Proceedings of the 33rd Annual Conference of the Cognitive Science Society*, eds. L. Carlson, C. Holscher, and T. Shipley, p. 3444. Cognitive Science Society, Austin, TX.

Jorde, L.B., and Wooding, S.P. 2004. "Genetic variation, classification and 'race.'" *Nature Genetics* **36**, S28.

Judson, O.P., and Normark, B.B. 1996. "Ancient asexual scandals." *Trends in Ecology & Evolution* **11**, A41.

Jung, C.G. 1971. *Psychological Types. Volume 6 of the Collected Works of C.G. Jung.* Princeton University Press, Princeton, NJ.

Kamenetz, A. 2015. *The Test: Why Our Schools Are Obsessed with Standardized Testing—But You Don't Have to Be.* PublicAffairs, New York.

Kandel, E.R., Schwartz, J.H., and Jessell, T.M. 2013. *Principles of Neural Science.* McGraw-Hill, New York.

Kane, M.J., Brown, L.H., McVay, J.C., Silvia, P.J., Myin-Germeys, I., and Kwapil, T.R. 2007. "For whom the mind wanders, and when—An experience-sampling study of working memory and executive control in daily life." *Psychological Science* **18**, 614.

Kauffman, S., and Levin, S. 1987. "Towards a general theory of adaptive walks on rugged landscapes." *Journal of Theoretical Biology* **128**, 11.

Kaufmann, T.D. 2004. *Toward a Geography of Art*. University of Chicago Press, Chicago.

Keane, M.A., Koza, J.R., and Streeter, M.J. 2005. "Apparatus for improved general purpose PID and non-PID controllers." (U.S. patent 6,847,851).

Keats, J. 2006. "John Koza has built an invention machine." *Popular Science*. April. http://www.popsci.com/scitech/article/2006-04/john -koza-has-built-invention-machine.

Keller, L.F. 1998. "Inbreeding and its fitness effects in an insular population of song sparrows (Melospiza melodia)." *Evolution* **52**, 240.

Keller, L.F., Arcese, P., Smith, J.N.M., Hochachka, W.M., and Stearns, S.C. 1994. "Selection against inbred song sparrows during a natural population bottleneck." *Nature* **372**, 356.

Kelley, T. 2001. *The Art of Innovation: Lessons in Creativity from IDEO, America's Leading Design Firm*. Doubleday, New York.

Kent, G.H., and Rosanoff, A.J. 1910. "A study of association in insanity." *American Journal of Psychiatry* **67**, 317.

Kettlewell, H.B.D. 1973. *The Evolution of Melanism: The Study of a Recurring Necessity*. Blackwell, Oxford, UK.

Killingsworth, M.A., and Gilbert, D.T. 2010. "A wandering mind is an unhappy mind." *Science* **330**, 932.

Kim, K.H. 2006. "Can we trust creativity tests? A review of the Torrance tests of Creative Thinking (TTCT)." *Creativity Research Journal* **18**, 3.

———. 2011. "The creativity crisis: The decrease in creative thinking scores on the Torrance tests of creative thinking." *Creativity Research Journal* **23**, 285.

Knapp, S., and Mallet, J. 2003. "Refuting refugia?" *Science* **300**, 71.

Koestler, A. 1964. *The Act of Creation*. MacMillan, New York.

Kohn, D. 2015. "Let the kids learn through play." *New York Times*. May 16.

Koo, S.-W. 2014. "An assault upon our children." *New York Times*. August 1.

Koza, J.R. 1992. *Genetic Programming: On the Programming of Computers by Means of Natural Selection*. MIT Press, Cambridge, MA.

Koza, J.R., Bennett III, F.H., Andre, D., and Keane, M.A. 1999. "The design of analog circuits by means of genetic programming." In *Evolutionary Design by Computers*, ed. P.J. Bentley, p. 365. Morgan Kaufman, San Francisco.

Koza, J.R., Keane, M.A., and Streeter, M.J. 2003. "Evolving inventions." *Scientific American* **288**, 52.

Kroto, H. 1988. "Space, stars, C-60, and soot." *Science* **242**, 1139.

Kroto, H.W., Heath, J.R., O'Brien, S.C., Curl, R.F., and Smalley, R.E. 1985. "C-60—Buckminsterfullerene." *Nature* **318**, 162.

Kroto, H., Heath, J., O'Brien, S., Curl, R., and Smalley, R. 1987. "Long carbon chain molecules in circumstellar shells." *The Astrophysical Journal* **314**, 352.

Kubler, G. 1962. *The Shape of Time*. Yale University Press, New Haven, CT.

Lamichhaney, S., Berglund, J., Almen, M.S., Maqbool, K., Grabherr, M., Martinez-Barrio, A., Promerova, M., Rubin, C.J., Wang, C., Zamani, N., Grant, B.R., Grant, P.R., Webster, M.T., and Andersson, L. 2015. "Evolution of Darwin's finches and their beaks revealed by genome sequencing." *Nature* **518**, 371.

Lamichhaney, S., Han, F., Webster, M.T., Andersson, L., Grant, B.R., and Grant, P.R. 2018. "Rapid hybrid speciation in Darwin's finches." *Science* **359**, 224.

Landauer, T.K., and Dumais, S.T. 1997. "A solution to Plato's problem: The latent semantic analysis theory of acquisition, induction, and representation of knowledge." *Psychological Review* **104**, 211.

Lariviere, V., Gingras, Y., and Archambault, E. 2009. "The decline in the concentration of citations, 1900–2007." *Journal of the American Society for Information Science and Technology* **60**, 858.

Larmer, B. 2014. "Inside a Chinese test-prep factory." *New York Times*. December 31.

Lee, F.S., Pham, X., and Gu, G. 2013. "The UK research assessment exercise and the narrowing of UK economics." *Cambridge Journal of Economics* **37**, 693.

Lee, S.S. 2013. "South Korea's dreaded college entrance is the stuff of high school nightmares, but is it producing 'robots'?" *CBS News*. November 7.

Leung, A.K.-Y., Maddux, W.W., Galinsky, A.D., and Chiu, C.-Y. 2008. "Multicultural experience enhances creativity: The when and how." *American Psychologist* **63**, 169.

Levin, S.R. 1982. "Aristotle's theory of metaphor." *Philosophy and Rhetoric* **15**, 24.

Levy, S. 2012. "Can an algorithm write a better news story than a human reporter?" *WIRED*. April 24.

Libbrecht, K.G. 2005. "The physics of snow crystals." *Reports on Progress in Physics* **68**, 855.

Lieberman, D., Tooby, J., and Cosmides, L. 2003. "Does morality have a biological basis? An empirical test of the factors governing moral sentiments relating to incest." *Proceedings of the Royal Society B-Biological Sciences* **270**, 819.

Liu, W.Y., Lin, C.C., Chiu, C.R., Tsao, Y.S., and Wang, Q.W. 2014. "Minimizing the carbon footprint for the time-dependent heterogeneous-fleet vehicle routing problem with alternative paths." *Sustainability* **6**, 4658.

Lohr, S. 2007. "John W. Backus, 82, Fortran developer, dies." *New York Times*. March 20.

Lubow, R.E. 1973. "Latent inhibition." *Psychological Bulletin* **79**, 398.

Lubow, R.E., and Gewirtz, J.C. 1995. "Latent inhibition in humans—data, theory, and implications for schizophrenia." *Psychological Bulletin* **117**, 87.

Lubow, R.E., Ingbergsachs, Y., Zalsteinorda, N., and Gewirtz, J.C. 1992. "Latent inhibition in low and high psychotic-prone normal subjects." *Personality and Individual Differences* **13**, 563.

Lynch, M. 2006. "The origins of eukaryotic gene structure." *Molecular Biology and Evolution* **23**, 450.

———. 2007. *The origins of genome architecture*. Sinauer Associates, Sunderland, MA.

Lynch, M., and Conery, J.S. 2000. "The evolutionary fate and consequences of duplicate genes." *Science* **290**, 1151.

———. 2003. "The origins of genome complexity." *Science* **302**.

MacFadden, B.J. 2005. "Fossil horses—evidence for evolution." *Science* **307**, 1728.

Mackenzie, S.M., Brooker, M.R., Gill, T.R., Cox, G.B., Howells, A.J., and Ewart, G.D. 1999. "Mutations in the white gene of Drosophila melanogaster affecting ABC transporters that determine eye colouration." *Biochimica et Biophysica Acta-Biomembranes* **1419**, 173.

Maddux, W.W., Adam, H., and Galinsky, A.D. 2010. "When in Rome...Learn why the Romans do what they do: How multicultural learning experiences facilitate creativity." *Personality and Social Psychology Bulletin* **36**, 731.

Maddux, W.W., and Galinsky, A.D. 2009. "Cultural borders and mental barriers: The relationship between living abroad and creativity." *Journal of Personality and Social Psychology* **96**, 1047.

Maeda, K., Nojiri, H., Shintani, M., Yoshida, T., Habe, H., and Omori, T. 2003. "Complete nucleotide sequence of carbazole/dioxin-degrading plasmid pCAR1 in Pseudomonas resinovorans strain CA10 indicates its mosaicity and the presence of large catabolic transposon Tn4676." *Journal of Molecular Biology* **326**, 21.

Majerus, M.E.N. 1998. *Melanism: Evolution in Action*. Oxford University Press, Oxford, UK.

Marcon, R.A. 2002. "Moving up the grades: Relationship between preschool model and later school success." *Early Childhood Research & Practice* **4**, 1.

Markus, H.R., and Kitayama, S. 1991. "Culture and the self—implications for cognition, emotion, and motivation." *Psychological Review* **98**, 224.

Martin, C. 2014. "Wearing your failures on your sleeve." *New York Times*. November 8.

Martin, O.C., and Wagner, A. 2009. "Effects of recombination on complex regulatory circuits." *Genetics* **183**, 673.

Martin, P. 2002. *Counting Sheep: The Science and Pleasures of Sleep and Dreams*. Harper Collins, London, UK.

Matai, R., Singh, S.P., and Mittal, M.L. 2010. "Traveling salesman problem: An overview of applications, formulations, and solution approaches." In *Traveling Salesman Problem, Theory, and Applications*, ed. D. Davendra. InTech, Rijeka, Croatia.

Maynard-Smith, J. 1970. "Natural selection and the concept of a protein space." *Nature* **255**, 563.

Maynard-Smith, J., Burian, R., Kauffman, S., Alberch, P., Campbell, J., Goodwin, B., Lande, R., Raup, D., and Wolpert, L. 1985.

"Developmental constraints and evolution." *Quarterly Review of Biology* **60**, 265.

McArdle, M. 2014. *The Up Side of Down: Why Failing Well Is the Key to Success*. Penguin, New York.

McGhee, G.R. 2007. *The Geometry of Evolution: Adaptive Landscapes and Theoretical Morphospaces*. Cambridge University Press, Cambridge, UK.

Mednick, S.A. 1962. "The associative basis of the creative process." *Psychological Review* **69**, 220.

Meng, G.N., Arkus, N., Brenner, M.P., and Manoharan, V.N. 2010. "The free-energy landscape of clusters of attractive hard spheres." *Science* **327**, 560.

Michaelian, K., Rendon, N., and Garzon, I.L. 1999. "Structure and energetics of Ni, Ag, and Au nanoclusters." *Physical Review B* **60**, 2000.

Miranda-Rottmann, S., Kozlov, A.S., and Hudspeth, A.J. 2010. "Highly specific alternative splicing of transcripts encoding BK channels in the chicken's cochlea is a minor determinant of the tonotopic gradient." *Molecular and Cellular Biology* **30**, 3646.

Mitchell, M. 1998. *An Introduction to Genetic Algorithms*. MIT Press, Cambridge, MA.

Montgomery, S.L. 1983. "Carnivorous caterpillars: The behavior, biogeography and conservation of Eupithecia (Lepidoptera: Geometridae) in the Hawaiian islands." *GeoJournal* **7**, 549.

Mooneyham, B.W., and Schooler, J.W. 2013. "The costs and benefits of mind-wandering: A review." *Canadian Journal of Experimental Psychology—Revue Canadienne De Psychologie Experimentale* **67**, 11.

Moore, C., and Mertens, S. 2011. *The Nature of Computation*. Oxford University Press, Oxford, UK.

"Morally bankrupt." 2005. *Economist*. April 15.

Mrazek, M.D., Franklin, M.S., Phillips, D.T., Baird, B., and Schooler, J.W. 2013. "Mindfulness training improves working memory capacity and GRE performance while reducing mind wandering." *Psychological Science* **24**, 776.

Mukherjee, S., Berger, M.F., Jona, G., Wang, X.S., Muzzey, D., Snyder, M., Young, R.A., and Bulyk, M.L. 2004. "Rapid analysis of the

DNA-binding specificities of transcription factors with DNA microarrays." *Nature Genetics* **36**, 1331.

Muscutt, K. 2007. "Composing with algorithms: An interview with David Cope." *Computer Music Journal* **31**, 10.

National Association for College Admission Counseling. 2008. "Report of the commission on the use of standardized tests in undergraduate admissions." Available at: http://www.nacacnet.org/research/PublicationsResources/Marketplace/research/Pages/TestingCommissionReport.aspx.

National Institutes of Health. 2012. "Biomedical research workforce working group report." Bethesda, MD.

National Science Foundation. 2014. "Report to the National Science Board on the National Science Foundation's merit review process. Fiscal Year 2013." Washington, DC.

Nemeth, C.J., and Kwan, J.L. 1987. "Minority influence, divergent thinking, and detection of correct solutions." *Journal of Applied Social Psychology* **17**, 788.

Ness, J., Welch, M., Giver, L., Bueno, M., Cherry, J., Borchert, T., Stemmer, W., and Minshull, J. 1999. "DNA shuffling of subgenomic sequences of subtilisin." *Nature Biotechnology* **17**, 893.

Newman, M.E.J., Strogatz, S.H., and Watts, D.J. 2001. "Random graphs with arbitrary degree distributions and their applications." *Physical Review E* **64**.

Niu, W.H., and Sternberg, R.J. 2001. "Cultural influences on artistic creativity and its evaluation." *International Journal of Psychology* **36**, 225.

———. 2003. "Societal and school influences on student creativity: The case of China." *Psychology in the Schools* **40**, 103.

Nocera, J. 2015. "How to grade a teacher." *New York Times*. June 16.

Normile, D. 2015. "Japan looks to instill global mindset in grads." *Science* **347**, 937.

Norris, L.C., Main, B.J., Lee, Y., Collier, T.C., Fofana, A., Cornel, A.J., and Lanzaro, G.C. 2015. "Adaptive introgression in an African malaria mosquito coincident with the increased usage of insecticide-treated bed nets." *Proceedings of the National Academy of Sciences of the United States of America* **112**, 815.

Norton, J.D. 2012. "Chasing the light. Einstein's most famous thought experiment." In *Thought Experiments in Philosophy, Science, and the Arts*, eds. J.R. Brown, M. Frappier, and L. Meynell, p. 123. Routledge, New York.

Oliver-Meseguer, J., Cabrero-Antonino, J.R., Dominguez, I., Leyva-Perez, A., and Corma, A. 2012. "Small gold clusters formed in solution give reaction turnover numbers of 10^7 at room temperature." *Science* **338**, 1452.

Osepchuk, J.M. 1984. "A history of microwave heating applications." *IEEE Transactions on Microwave Theory and Techniques* **32**, 1200.

Pachet, F. 2008. "The future of content is in ourselves." ACM *Journal of Computers in Entertainment* **6**, 1.

———. 2012. "Musical virtuosity and creativity." In *Computers and Creativity*, eds. J. McCormack and M. d'Inverno, p. 115. Springer, Berlin, Germany.

Padel, R. 2008. *The Poem and the Journey: 60 Poems for the Journey of Life*. Vintage Books, New York.

Palmer, S.E. 1999. *Vision Science*. MIT Press, Cambridge, MA.

Pappano, L. 2014. "Learning to think outside the box." *New York Times*. February 5.

Pavitt, K. 2001. "Public policies to support basic research: What can the rest of the world learn from US theory and practice? (And what they should not learn)." *Industrial and Corporate Change* **10**, 761.

Pegadaraju, V., Nipper, R., Hulke, B., Qi, L.L., and Schultz, Q. 2013. "De novo sequencing of sunflower genome for SNP discovery using RAD (Restriction site Associated DNA) approach." *BMC Genomics* **14**.

Pellegrini, A.D. 1988. "Elementary school childrens' rough-and-tumble play and social competence." *Developmental Psychology* **24**, 802.

Pennisi, E. 2016. "Shaking up the tree of life." *Science* **354**, 817.

Pigliucci, M. 2012. "Landscapes, surfaces, and morphospaces: What are they good for?" In *The Adaptive Landscape in Evolutionary Biology*, eds. E.I. Svensson and R. Calsbeek, p. 26. Oxford University Press, Oxford, UK.

Piketty, T. 2014. *Capital in the Twenty-first Century*. Belknap Press, Cambridge, MA.

Pinker, S. 2007. *The Stuff of Thought*. Penguin, New York.

Plotkin, R. 2009. *The Genie in the Machine*. Stanford University Press, Stanford, CA.

Plucker, J.A. 1999. "Is the proof in the pudding? Reanalyses of Torrance's (1958 to present) longitudinal data." *Creativity Research Journal* 12, 103.

Plunkett, R.J. 1986. "The history of polytetrafluoroethylene: Discovery and development." In *High Performance Polymers: Their Origin and Development. Proceedings of the Symposium on the History of High Performance Polymers at the American Chemical Society Meeting*, eds. R.B. Seymour and G.S. Kirshenbaum, p. 261. Elsevier, New York.

Podolny, S. 2015. "If an algorithm wrote this, how would you even know?" *New York Times*. March 8.

Pomerantz, E.M., Ng, F.F.Y., Cheung, C.S.S., and Qu, Y. 2014. "Raising happy children who succeed in school: Lessons from China and the United States." *Child Development Perspectives* 8, 71.

Porter, E. 2015. "American innovation lies on a weak foundation." *New York Times*. May 19.

Preston, J. 2015a. "In turnabout, Disney cancels tech worker layoffs." *New York Times*. June 16.

———. 2015b. "Last task after layoff at Disney: Train foreign replacements." *New York Times*. June 3.

Prins, C. 2004. "A simple and effective evolutionary algorithm for the vehicle routing problem." *Computers & Operations Research* 31, 1985.

Provine, W.B. 1986. *Sewall Wright and Evolutionary Biology*. University of Chicago Press, Chicago.

Pruitt, J.N., and Riechert, S.E. 2011. "Nonconceptive sexual experience diminishes individuals' latency to mate and increases maternal investment." *Animal Behaviour* 81, 789.

Pusey, A., and Wolf, M. 1996. "Inbreeding avoidance in animals." *Trends in Ecology & Evolution* 11, 201.

Pusey, A.E., and Packer, C. 1987. "The evolution of sex-biased dispersal in lions." *Behaviour* 101, 275.

Raillard, S., Krebber, A., Chen, Y.C., Ness, J.E., Bermudez, E., Trinidad, R., Fullem, R., Davis, C., Welch, M., Seffernick, J., Wackett, L.P., Stemmer, W.P.C., and Minshull, J. 2001. "Novel enzyme activities and functional plasticity revealed by recombining highly homologous enzymes." *Chemistry & Biology* **8**, 891.

Raina, M. 1975. "Parental perception about ideal child: A cross-cultural study." *Journal of Marriage and the Family* **37**, 229.

Raman, K., and Wagner, A. 2011. "The evolvability of programmable hardware." *Journal of the Royal Society Interface* **8**, 269.

Ramon y Cajal, S. 1951. *Precepts and Counsels on Scientific Investigation: Stimulants of the Spirit. (Translated by Sanchez-Perez, J.S.)*. Pacific Press Publishing Association, Mountain View, CA.

Raup, D.M. 1967. "Geometric analysis of shell coiling: Coiling in ammonoids." *Journal of Paleontology* **41**, 43.

Rechenberg, I. 1973. *Evolutionsstrategie*. Frommann-Holzboog, Stuttgart, Germany.

Rees, J. 2010. *Künstler auf Reisen*. Wissenschaftliche Buchgesellschaft, Darmstadt, Germany.

Rhodes, G. 1999. *Crystallography Made Crystal Clear*. Academic Press, San Diego, CA.

Rich, M. 2015. "Out of the books in kindergarten, and into the sandbox." *New York Times*. June 9.

Rieseberg, L.H., Kim, S.C., Randell, R.A., Whitney, K.D., Gross, B.L., Lexer, C., and Clay, K. 2007. "Hybridization and the colonization of novel habitats by annual sunflowers." *Genetica* **129**, 149.

Robinson, K.M., Sieber, K.B., and Hotopp, J.C.D. 2013. "A review of bacteria-animal lateral gene transfer may inform our understanding of diseases like cancer." *PLoS Genetics* **9**.

Root-Bernstein, R., Allen, L., Beach, L., Bhadula, R., Fast, J., Hosey, C., Kremkow, B., Lapp, J., Lonc, K., Pawelec, K., Podufaly, A., Russ, C., Tennant, L., Vrtis, E., and Weinlander, S. 2008. "Arts foster scientific success: Avocations of Nobel, National Academy, Royal Society, and Sigma Xi members." *Journal of Psychology of Science and Technology* **1**, 51.

Root-Bernstein, R.S., and Root-Bernstein, M. 1999. *Sparks of Genius*. Houghton Mifflin, New York.

Ropars, J., de la Vega, R.C.R., Lopez-Villavicencio, M., Gouzy, J., Sallet, E., Dumas, E., Lacoste, S., Debuchy, R., Dupont, J., Branca, A., and Giraud, T. 2015. "Adaptive horizontal gene transfers between multiple cheese-associated fungi." *Current Biology* **25**, 2562.

Rosen, W. 2010. *The Most Powerful Idea in the World*. University of Chicago Press, Chicago.

Rosenberg, N., and Nelson, R.R. 1994. "American universities and technical advance in industry." *Research Policy* **23**, 323.

Rothenberg, A. 1976. "Homospatial thinking in creativity." *Archives of General Psychiatry* **33**, 17.

———. 1980. "Visual art—homospatial thinking in the creative process." *Leonardo* **13**, 17.

———. 1986. "Artistic creation as stimulated by superimposed versus combined composite visual images." *Journal of Personality and Social Psychology* **50**, 370.

———. 1995. "Creative cognitive-processes in Kekule's discovery of the structure of the benzene molecule." *American Journal of Psychology* **108**, 419.

———. 2015. *Flight from Wonder*. Oxford University Press, Oxford, UK.

Ruef, K. 2005. "Research basis of the Private Eye." http://www.the-private-eye.com/pdfs/ResearchBasis.pdf.

Runco, M.A. 1992. "Children's divergent thinking and creative ideation." *Developmental Review* **12**, 233.

———. 2001. "Creativity training." In *International Encyclopedia of the Social & Behavioral Sciences*, eds. N.J. Smelser and P.B. Baltes, p. 2900. Elsevier, Oxford, UK.

———. 2014. *Creativity: Theories and Themes: Research, Development, and Practice*. Academic Press, London.

Russell, R.J., Scott, C., Jackson, C.J., Pandey, R., Pandey, G., Taylor, M.C., Coppin, C.W., Liu, J.W., and Oakeshott, J.G. 2011. "The evolution of new enzyme function: Lessons from xenobiotic metabolizing bacteria versus insecticide-resistant insects." *Evolutionary Applications* **4**, 225.

Sahlberg, P. 2015. *Finnish Lessons 2.0*. Teachers College Press, New York.

Salter, A.J., and Martin, B.R. 2001. "The economic benefits of publicly funded basic research: A critical review." *Research Policy* **30**, 509.

Saunders, W.B., Work, D.M., and Nikolaeva, S.V. 2004. "The evolutionary history of shell geometry in Paleozoic ammonoids." *Paleobiology* **30**, 19.

Sawyer, K. 2013. *Zig-zag: The Surprising Path to Greater Creativity*. Jossey-Bass, San Francisco.

Schiappa, J., and Van Hee, R. 2012. "From ants to staples: History and ideas concerning suturing techniques." *Acta Chirurgica Belgica* **112**, 395.

Schmidt, M., and Lipson, H. 2009. "Distilling free-form natural laws from experimental data." *Science* **324**, 81.

Schooler, J.W., Mrazek, M.D., Franklin, M.S., Baird, B., Mooneyham, B.W., Zedelius, C., and Broadway, J.M. 2014. "The middle way: Finding the balance between mindfulness and mind-wandering." *Psychology of Learning and Motivation* **60**, 1.

Schwartz, T. 2013. "Relax! You'll be more productive." *New York Times*. February 9.

Scott, G., Leritz, L.E., and Mumford, M.D. 2004. "The effectiveness of creativity training: A quantitative review." *Creativity Research Journal* **16**, 361.

Scott, R.A. 2003. *The Gothic Enterprise*. University of California Press, Berkeley.

Sessa, B. 2008. "Is it time to revisit the role of psychedelic drugs in enhancing human creativity?" *Journal of Psychopharmacology* **22**, 821.

Shepher, J. 1971. "Mate selection among second generation Kibbutz adolescents and adults—incest avoidance and negative imprinting." *Archives of Sexual Behavior* **1**, 293.

Sibani, P., Schon, J.C., Salamon, P., and Andersson, J.O. 1993. "Emergent hierarchical structures in complex-system dynamics." *Europhysics Letters* **22**, 479.

Simonton, D.K. 1975. "Age and literary creativity—cross-cultural and transhistorical survey." *Journal of Cross-Cultural Psychology* **6**, 259.

———. 1977. "Creative productivity, age, and stress—biographical time-series analysis of 10 classical composers." *Journal of Personality and Social Psychology* **35**, 791.

———. 1985. "Quality, quantity, and age—the careers of ten distinguished psychologists." *International Journal of Aging & Human Development* **21**, 241.

———. 1988. *Scientific Genius*. Cambridge University Press, New York.

———. 1990. "Political pathology and societal creativity." *Creativity Research Journal* **3**, 85.

———. 1994. *Greatness: Who Makes History and Why*. Guilford Press, New York.

———. 1997. "Foreign influence and national achievement: The impact of open milieus on Japanese civilization." *Journal of Personality and Social Psychology* **72**, 86.

———. 1999. *Origins of Genius: Darwinian Perspectives on Creativity*. Oxford University Press, New York.

———. 2007a. "The creative process in Picasso's Guernica sketches: Monotonic improvements versus nonmonotonic variants." *Creativity Research Journal* **19**, 329.

———. 2007b. "Picasso's Guernica creativity as a Darwinian process: Definitions, clarifications, misconceptions, and applications." *Creativity Research Journal* **19**, 381.

———. 2014. "The mad-genius paradox: Can creative people be more mentally healthy but highly creative people more mentally ill?" *Perspectives on Psychological Science* **9**, 470.

Simpson, G.G. 1944. *Tempo and Mode in Evolution*. Hafner, New York.

Sinatra, R., Wang, D., Deville, P., Song, C., and Barabasi, A.L. 2016. "Quantifying the evolution of individual scientific impact." *Science* **354**, 596.

Skipper Jr., R.A., and Dietrich, M.R. 2012. "Sewall Wright's adaptive landscape: Philosophical reflections on heuristic value." In *The Adaptive Landscape in Evolutionary Biology*, eds. E.I. Svensson and R. Calsbeek, p. 16. Oxford University Press, Oxford, UK.

Slack, C. 2002. *Noble Obsession: Charles Goodyear, Thomas Hancock, and the Race to Unlock the Greatest Industrial Secret of the Nineteenth Century*. Hyperion, New York.

Smalley, R.E. 1992. "Self-assembly of the fullerenes." *Accounts of Chemical Research* **25**, 98.

Smith, D.C. 1991. "Why do so few animals form endosymbiotic associations with photosynthetic microbes?" *Philosophical Transactions of the Royal Society of London Series B-Biological Sciences* **333**, 225.

Smith, S. 2013. "Iamus: Is this the 21st century's answer to Mozart?" *BBC News Technology*. January 3.

Sobel, R.S., and Rothenberg, A. 1980. "Artistic creation as stimulated by superimposed versus separated visual images." *Journal of Personality and Social Psychology* **39**, 953.

Spinka, M., Newberry, R.C., and Bekoff, M. 2001. "Mammalian play: Training for the unexpected." *Quarterly Review of Biology* **76**, 141.

State Secretariat for Education and Research. 2011. "Higher education and research in Switzerland." Bern, Switzerland. Available at: https://www.sbfi.admin.ch/dam/sbfi/en/dokumente/hochschulen _und_forschunginderschweiz.pdf.download.pdf/higher_education andresearchinswitzerland.pdf.

Stemmer, W. 1994. "DNA shuffling by random fragmentation and reassembly—in-vitro recombination for molecular evolution." *Proceedings of the National Academy of Sciences of the U.S.A.* **91**, 10747.

Stern, N. 1978. "Age and achievement in mathematics: A case study in the sociology of science." *Social Studies of Science* **8**, 127.

Stewart, J.B. 2015. "A fearless culture fuels U.S. tech giants." *New York Times*. June 18.

Sugimoto, C.R., Robinson-Garcia, N., Murray, D.S., Yegros-Yegros, A., Costas, R., and Larivière, V. 2017. "Scientists have most impact when they're free to move." *Nature News* **550**, 29.

Sulloway, F.J. 1982. "Darwin and his finches: The evolution of a legend." *Journal of the History of Biology* **15**, 1.

Sun, H.P., Lin, Y., and Pan, C.W. 2014. "Iris color and associated pathological ocular complications: A review of epidemiologic studies." *International Journal of Ophthalmology* **7**, 872.

Szendro, I.G., Schenk, M.F., Franke, J., Krug, J., and de Visser, J.A.G.M. 2013. "Quantitative analyses of empirical fitness landscapes." *Journal of Statistical Mechanics: Theory and Experiment* **2013**, P01005.

Szulkin, M., Stopher, K.V., Pemberton, J.M., and Reid, J.M. 2013. "Inbreeding avoidance, tolerance, or preference in animals?" *Trends in Ecology & Evolution* **28**, 205.

Talhelm, T., Zhang, X., Oishi, S., Shimin, C., Duan, D., Lan, X., and Kitayama, S. 2014. "Large-scale psychological differences within China explained by rice versus wheat agriculture." *Science* **344**, 603.

Taylor, R.P., Spehar, B., Donkelaar, P.V., and Hagerhall, C.M. 2011. "Perceptual and physiological responses to Jackson Pollock's fractals." *Frontiers in Human Neuroscience* **5**, 1.

"Test-taking in South Korea: Point me at the SKY." 2013. *Economist.* November 8.

Torrance, E.P. 1966. *The Torrance Tests of Creative Thinking—Norms—Technical Manual Research Edition—Verbal Tests, Forms A and B—Figural Tests, Forms A and B.* Personnel Press, Princeton, NJ.

———. 1972. "Can we teach children to think creatively?" *Journal of Creative Behavior* **6**, 114.

———. 1988. "The nature of creativity as manifest in its testing." In *The Nature of Creativity*, ed. R.J. Sternberg, p. 43. Cambridge University Press, Cambridge, UK.

Tourangeau, R., and Rips, L. 1991. "Interpreting and evaluating metaphors." *Journal of Memory and Language* **30**, 452.

Turing, A.M. 2013. "Computing machinery and intelligence." *Mind* **LIX**, 433.

Upmanyu, V.V., Bhardwaj, S., and Singh, S. 1996. "Word-association emotional indicators: Associations with anxiety, psychoticism, neuroticism, extraversion, and creativity." *Journal of Social Psychology* **136**, 521.

Van Tonder, G.J., Lyons, M.J., and Ejima, Y. 2002. "Perception psychology: Visual structure of a Japanese Zen garden." *Nature* **419**, 359.

Velasquez-Manoff, M. 2017. "What biracial people know." *New York Times*. March 4.

Verde, T. 2012. "The point of the arch." *Aramco World*. May/June, http://archive.aramcoworld.com/issue/201203/the.point.of.the.arch.htm.

von Helmholtz, H. 1908. "An autobiographical sketch." In *Popular Lectures on Scientific Subjects, Second Series* (translated by E.Atkinson), p. 266. Longmans, Green, and Co., London.

Wagner, A. 2005. "Energy constraints on the evolution of gene expression." *Molecular Biology and Evolution* **22**, 1365.

———. 2007. "Energy costs constrain the evolution of gene expression." *Journal of Experimental Zoology Part B-Molecular and Developmental Evolution* **308B**, 322.

———. 2012. "The role of randomness in Darwinian Evolution." *Philosophy of Science* **79**, 95.

———. 2014. *Arrival of the Fittest: Solving Evolution's Greatest Puzzle.* Current, New York.

Wagner, C.S., and Jonkers, K. 2017. "Open countries have strong science." *Nature* **550**, 32.

Wales, D.J. 2003. *Energy Landscapes.* Cambridge University Press, Cambridge, UK.

Walworth, C. 2015. "Paly school board rep: 'The sorrows of young Palo Altans.'" *Palo Alto Online.* https://www.paloaltoonline.com /news/2015/03/25/guest-opinion-the-sorrows-of-young-palo-altans.

Wang, C., Yu, S.C., Chen, W., and Sun, C. 2013. "Highly efficient light-trapping structure design inspired by natural evolution." *Scientific Reports* **3**.

Watson, J.D., and Crick, F.H. 1953. "A structure for deoxyribose nucleic acids." *Nature* **171**, 737.

Weinreich, D.M., Delaney, N.F., DePristo, M.A., and Hartl, D.L. 2006. "Darwinian evolution can follow only very few mutational paths to fitter proteins." *Science* **312**, 111.

Weirauch, M.T., Yang, A., Albu, M., Cote, A.G., Montenegro-Montero, A., Drewe, P., Najafabadi, H.S., Lambert, S.A., Mann, I., Cook, K., Zheng, H., Goity, A., van Bakel, H., Lozano, J.-C., Galli, M., Lewsey, M.G., Huang, E., Mukherjee, T., Chen, X., Reece-Hoyes, J.S., Govindarajan, S., Shaulsky, G., Walhout, A.J.M., Bouget, F.-Y., Ratsch, G., Larrondo, L.F., Ecker, J.R., and Hughes, T.R. 2014. "Determination and inference of eukaryotic transcription factor sequence specificity." *Cell* **158**, 1431.

Weisberg, R.W. 2004. "On structure in the creative process: A quantitative case-study of the creation of Picasso's Guernica." *Empirical Studies of the Arts* **22**, 23.

Weisberg, R.W., and Hass, R. 2007. "We are all partly right: Comment on Simonton." *Creativity Research Journal* **19**, 345.

Wenner, M. 2009. "The serious need for play." *Scientific American Mind.* January 28.

West, G.B., Brown, J.H., and Enquist, B.J. 1997. "A general model for the origin of allometric scaling laws in biology." *Science* **276**, 122.

Westby, E.L., and Dawson, V.L. 1995. "Creativity: Asset or burden in the classroom?" *Creativity Research Journal* **8**, 1.

Whittaker, R.J., and Fernandez-Palacios, J.M. 2007. *Island Biogeography*. Oxford University Press, Oxford, UK.

Wilson, R.R. 1992. "Starting Fermilab." Fermilab. Retrieved August 20, 2014, from http://history.fnal.gov/GoldenBooks/gb_wilson2.html.

Wright, S. 1932. "The roles of mutation, inbreeding, crossbreeding, and selection in evolution." *Proceedings of the Sixth International Congress of Genetics in Ithaca, New York* **1**, 356.

———. 1978. "The relation of livestock breeding to theories of evolution." *Journal of Animal Science* **46**, 1192.

Wu, N.C., Dai, L., Olson, C.A., Lloyd-Smith, J.O., and Sun, R. 2016. "Adaptation in protein fitness landscapes is facilitated by indirect paths." *Elife* **5**, e16965.

Wuchty, S., Jones, B.F., and Uzzi, B. 2007. "The increasing dominance of teams in production of knowledge." *Science* **316**, 1036.

Zappe, H. 2013. "Bridging the market gap." *Nature* **501**, 483.

Zeng, L.A., Proctor, R.W., and Salvendy, G. 2011. "Can traditional divergent thinking tests be trusted in measuring and predicting real-world creativity?" *Creativity Research Journal* **23**, 24.

Zha, P., Walezyk, J.J., Griffith-Ross, D.A., Tobacyk, J.J., and Walczyk, D.F. 2006. "The impact of culture and individualism-collectivism on the creative potential and achievement of American and Chinese adults." *Creativity Research Journal* **18**, 355.

Zhao, Y. 2014. *Who's Afraid of the Big Bad Dragon? Why China Has the Best (and Worst) Education System in the World*. Jossey-Bass, San Francisco.

Zhao, Y., and Gearin, B. 2016. "Squeezed out." In *Creative Intelligence in the 21st Century*, eds. D. Ambrose and R.J. Sternberg, p. 121. Sense Publications, Rotterdam.

Figure Credits

Figure 1. Used with permission from Shutterstock.

Figure 1.1. Line drawing by author. "*Biston betularia* 7200" and "*Betularia f. carbonaria* 7209" images by Olaf Leillinger (CC BY-SA 2.5).

Figure 1.2. Line drawing by author. "*Biston betularia* 7200" and "*Betularia f. carbonaria* 7209" images by Olaf Leillinger (CC BY-SA 2.5).

Figure 1.3. "*Dichotomosphinctes* fossil ammonite" by James St. John (CC BY 2.0); line drawings from "Figure 6" of Saunders et al. (2004), *Paleobiology* 30, 19–43.

Figure 1.4. Created by author.

Figure 1.5. Line drawing by author. "*Biston betularia* 7200" and "*Betularia f. carbonaria* 7209" images by Olaf Leillinger (CC BY-SA 2.5).

Figure 3.1. Created by author.

Figure 3.2. Created by author.

Figure 4.1. Created by author.

Figure 5.1. "C60a" by Michael Ströck (CC BY-SA 3.0).

Figure 5.2. Created by author.

Figure 6.1. Created by author.

Index

absolute zero, 110

Acemoglu, Daron, 215

adaptive landscapes

  altitude/elevation and, 31, 37

  complexity/possibilities and, 31–32, 37, 38–39, 52

  descriptions, 5, 5 (fig.), 17, 27, 32

  energy landscapes vs., 108, 109–110

  environmental changes and, 19

  horizontal/vertical dimensions meaning, 17

  importance (summary), 219–222

  location and, 31, 37

  microarray technology and, 50–51

  natural selection and, 18–19, 18 (fig.), 19 (fig.)

  numbers of dimensions and, 81–83, 82 (fig.)

  peak numbers/estimates and, 38–39

  peppered moth and, 17–19, 18 (fig.), 19 (fig.)

  recombination and, 7, 90–99, 140

  single-peaked vs. multi-peaked/ solutions and, 52

  two-peaked landscapes and, 20–21, 23–24, 23 (fig.)

  warning coloration and, 25–26

  Wright's "landscape concept" and, 5–6, 17, 20–21, 24, 28, 31, 32, 36, 37, 38, 40, 51–52

  See also creativity

adaptive landscapes/ high-dimensional

  connections/effects and, 83–88, 98–99

  human innovators/innovations and, 93–95

  laboratory experiments, 83–88

  metabolic machineries and, 86

  regulator proteins and, 85–86

adenine base, 34

"albino" seedlings, 58

algorithms

  definition, 116

  evolution analogy, 126, 127

  of evolution/harnessing and, 126–130

  greedy algorithm, 123, 124

  significance, 130

  types overview, 128–129

Andreas Wagner is a professor and chairman in the Department of Evolutionary Biology and Environmental Studies at the University of Zurich and an external professor at the Sante Fe Institute. He is also the author of four books on evolutionary innovation. He lives in Zurich, Switzerland.